Science in Culture
c. 350–c. 1750

Volume 1

Meditation, Invention, and Designing Thought in the Augustinian Middle Ages

Science in Culture c. 350–c. 1750

This series focuses on the history and culture of science globally between c.350 and c.1750. It considers all relevant thematic areas, including acculturation of scientific ideas and their adoption or appropriation in different cultural milieux, and aims to promote a wider consideration of human engagement with, and understanding of, natural phenomena. Such areas might include human-animal interaction, environmental history, astronomy, astrology and cosmology, medicine (human and animal), alchemy, and practical disciplines such as metallurgy as well as theoretical subjects.

While centred on the medieval and early modern eras, studies on earlier periods that look forward or those on later periods that look back are also considered, where appropriate. The geographical range is equally broad, encompassing all scientific cultures worldwide. The series encourages wider interdisciplinary approaches, including collaboration with modern scientists, as well as more traditional contextual studies, and is particularly interested in the possibilities offered by collaboration across areas of expertise which bring areas of specialism into dialogue.

The series welcomes standard monographs and coherent essay collections, as well as mid-range studies (ca. 45,000 words) and editions.

We are happy to consider both conventional and Open Access models of book publishing. Further information about the series and guidance on submitting a proposal can be found on the Durham University IMEMS Press website.

Other books in the series may be viewed on the Boydell & Brewer website.

Meditation, Invention, and Designing Thought in the Augustinian Middle Ages

Mary Carruthers

Durham University
IMEMS PRESS

First published 2025
A Durham University IMEMS Press publication
in association with
The Boydell Press
an imprint of Boydell & Brewer Ltd
PO Box 9, Woodbridge, Suffolk IP12 3DF, UK
and of Boydell & Brewer Inc.
668 Mt Hope Avenue, Rochester, NY 14620–2731, USA
website: www.boydellandbrewer.com
and with the
Institute of Medieval and Early Modern Studies
University of Durham

ISBN 978-1-914967-22-1

A CIP catalogue record for this book is available
from the British Library

Contents

Illustrations

Figures

Drawings

The author and publisher are grateful to all the institutions and individuals listed for permission to reproduce the materials in which they hold copyright. Every effort has been made to trace the copyright holders; apologies are offered for any omission, and the publisher will be pleased to add any necessary acknowledgement in subsequent editions.

Acknowledgments

This book develops from three different series of lectures that I have delivered over a period of a dozen years, and I particularly thank my hosts on those occasions. In the spring of 2012, I gave The Speaker's Lectures, a set of six (two of which I combined into a discussion-oriented seminar) at the invitation of the Faculty of Theology in Oxford. The next series was the four Clark Lectures in 2016, given in Cambridge at the invitation of the Master and Fellows of Trinity College, where, since I was in residence for six weeks, I was able to immerse myself in the manuscripts of the Wren Library as well as in many profitable conversations with friends, old and new. A year later, in the hospitable setting of the Schoenberg Library of the University of Pennsylvania, I gave the Rosenbach Lectures, sponsored by the Rosenbach Foundation in the Free Library of Philadelphia as well as the Schoenberg, where I was generously hosted by the Rare Books and Manuscripts staff headed by Dr. Lynn Ransom. All of these occasions were not only educational for me but often great fun, as my audiences generously questioned and commented on my thoughts, and we pooled our diverse knowledge. I like to think these were something like the colloquia of the canons of St. Victor, that delightfully fruitful group of scholars including both Hugh and Richard, whose works have taught me just how creative the figures of geometry can be for the thought of those who knew how best to use them in seeking to understand rationally not only the natural and social human worlds, the worlds of *scientia*, but *sapientia*, the divinely ordained cosmos itself.

I was in residence at Durham University's Institute for Advanced Studies for October–December 2010 (at the kind instigation of Corinne Saunders), and during that time I was invited to attend, as a fly-on-the-wall, a seminar in connection with this project, as also to attend more generally the activities of Durham's Institute for Medieval Studies (as it was then called). This experience, coming after two years as Chair of the English Department and then immediately a five-year term as Dean for Humanities at New York University, plus my welcome after that into the extraordinary community of medievalists working in the UK, gave me insights into what universities could do to support truly multidisciplinary research projects of exceptional worth as no single discipline, division, or indeed university, could. At Durham's Institute of Advanced Study I was invited to sit in on some of the seminars of the incipient "Ordered Universe" project, then a gleam in the eyes of a core of scholars. They were focused on translating, with extensive commentaries on its natural science, the difficult Latin text of Robert Grosseteste's *De colore*, a project requiring extensive knowledge of medieval history, Latin scientific texts, the physics, optics, and psychology of light, and our human perceptions of it as distinct colors. There I got to know Tom McLeish, Giles Gaspar, Faith Wallis, and Hannah Smithson, exemplary scholars whose own work demonstrates how the sciences and the humanities need one another in order to stay true as language, as history, as science, as ethics, as mathematics – indeed the whole of Grammatica-Logica and the Moral and Natural Sciences. I owe particular thanks to my friendship with Tom McLeish, who died too soon and suddenly to see this book finished, but who himself exemplified the kind of cross-disciplinary studies I was trying to undertake.

Soon after my time in Durham, I was invited as a short-term visitor to a seminar sponsored by the Israel Institute for Advanced Study at the Hebrew University of Jerusalem. The topic was "envisioning knowledge," and during the seven weeks I was in residence I profited greatly in developing my research on geometry, meditation, and rhetoric by conversations with many Israeli scholars whom I would never have encountered otherwise. Though the group centered on diagrams as objects of art history, what I learned there encouraged me to enlarge this subject into questions of envisioning for mystical contemplation and the role of geometry in enabling many sorts of inventive thinking. That time, plus my time in Durham with Robert Grosseteste and the Ordered Universe group, has brought this current book together.

It is often believed that humanities scholarship is a solitary activity. That has certainly never been my experience in researching any of my own works. Humanists do not work best in "teams" organized as labs or term-limited projects like those centered in the various non-teaching Institutes that have popped up in our universities in the past few decades. We rarely produce multi-authored essays. Discussion, conversation, constant questioning, reading one another's work – these are our vital collaborations, and any corporate agent or administrator who fails to grasp this has wandered off into a wilderness of error. For our work, teaching in seminar and small discussion groups is essential. Our collaborations are most often also with old books in libraries, for research libraries are the laboratories of scholars, where they meet not only their books but other scholars working on similar questions and problems. Humanities scholars learn from and with one another; this is the necessary role of tea-rooms, cafes, and other common spaces in and near libraries. My experiences in Durham and elsewhere have taught me also that natural and social scientists who would like to join these conversations are most welcome.

The Fellows of St Hilda's College, Oxford, offered me Visiting Fellow's status, IT support, and conversation for several fruitful years, as for one crucial year did the Fellows of Harris-Manchester College. All Souls College has enabled both continuing and changing exchanges, and was a welcome conversation and dining base even during the difficult lockdown years of the COVID pandemic. Oxford University also offers an unparalleled set of ongoing seminars and sponsored lecture series. In particular for this project, the seminars in Medieval Church and Culture, Medieval History, Medieval and Renaissance Music, and Medieval Visual Culture provided many stimulating sessions, even when the topic wasn't directly relevant to my current project. The Oxford Dante Society has afforded similar scholarly riches, always including dinner.

My particular thanks are to various friends upon whom I have prevailed for editorial and scholarly advice, and who have read and commented most helpfully on parts of this book as it made its slow progress, and often have also walked and talked and dined with me in the best monastic way: Margaret Bent, Paul Binski, Yossi Chayez, Michael Clanchy, Nicola Coldstream, Catherine Conybeare, Helen Cooper, Yadin Dudai, Kathy H. Eden, Ayelet Even-Ezra, David Ganz, Julian and Christa Gardner, Bob and Barbara Hanning, Martin Kauffmann, Marcia Kupfer, Elizabeth Eva Leach, Henrietta Leyser, Alastair Minnis, Jonathan Morton, Jennifer Rampling, Susan Rankin, Howard Robinson, Lucy Freeman Sandler, Hamish Scott, Julia Smith, Lesley Smith, Eric Southworth, Helen Swift, Andrew S. Wilson. Special thanks to Elly Truitt and to John N. Crossley, who read the entire manuscript and made most helpful comments before its submission. Thanks also to the series editorial group at Durham University IMEMS Press, in particular to Giles Gaspar, who never lost faith in this book, and to the editorial and production teams of Boydell & Brewer, in particular Elizabeth McDonald and Caitlin Smith. My particular thanks to Dr. B. A. Blackburn for preparing the Index.

Any of their advice not taken, and any errors I have made since are entirely of my own doing.

Mary Carruthers, Oxford and New York, January 2025

Abbreviations

Anticlaud.	Alan of Lille, *Anticlaudianus*, ed. W. Wetherbee.
CCCM	Corpus Christianorum, continuatio medievalis (Turnhout; online in LLT)
CCSL	Corpus Christianorum – Series Latina (Turnhout; online in LLT)
Consol.	Boethius, *Consolatio Philosophiae*, ed. L. Biehler.
CSEL	Corpus Scriptorum Ecclesiasticorum Latinorum
De contempl.	Richard of St. Victor, *De contemplatione maior*, ed. J. Grosfillier.
De Gen. ad litt.	Augustine, *De Genesi ad litteram*, ed. J. Zycha.
De nuptiis	Martianus Capella, *De nuptiis Philologiae et Mercurii*, ed. J. Willis.
DMLBS	Dictionary of Medieval Latin from British Sources (Brepols online database)
DOML	Dumbarton Oaks Medieval Library; Harvard University Press
DS	*Dictionnaire de Spiritualité*, ed. M. Villiers, 17 vols. (Paris, 1937–95)
Etymol.	Isidore of Seville, *Etymologiarum Sive Originum Libri XX*, ed. W. Lindsay.
FC	The Fathers of the Church series; Catholic University Press
Inst.	Cassiodorus, *Institutiones divinarum et saecularium litterarum*, ed. R. Mynors.
L&S	*A Latin Dictionary*, ed. C. Lewis and C. Short. Oxford University Press, 1998 (online)
LCL	Loeb Classical Library series; Harvard University Press.

Liddell & Scott	C. Liddell, R. Scott et al., *Greek–English Lexicon*. Oxford University Press, 1996 (online).
LLT	Library of Latin Texts (Brepols online database)
OCD	*Oxford Classical Dictionary*, 4th ed. (Oxford University Press, 2012)
OED	*Oxford English Dictionary* (unabridged), Oxford University Press (online)
OLD	*Oxford Latin Dictionary*, ed. P. Glare, 2nd ed. (Oxford University Press, 2012)
PG	*Patrologia Graeca*, ed. J-P. Migne (online database)
PL	*Patrologia Latina*, ed. J-P. Migne (online database)
SBO	*Sancti Bernardi Opera*, ed. C. Talbot, J. Leclercq, H. Rochais. 8 vols. Rome, 1957–87.

Figure 1 London, British Library, Cotton Tiberius C.VI, fol. 7v.
From the British Library Archive, reproduced by permission.

Figure 2 St. Gallen, Stiftsbibliothek, Cod. Sang. 817, p. 344.
Reproduced by permission.

Figure 3
Cambridge,
University
Library, Gg1.1,
fol. 490v. Public
Domain.

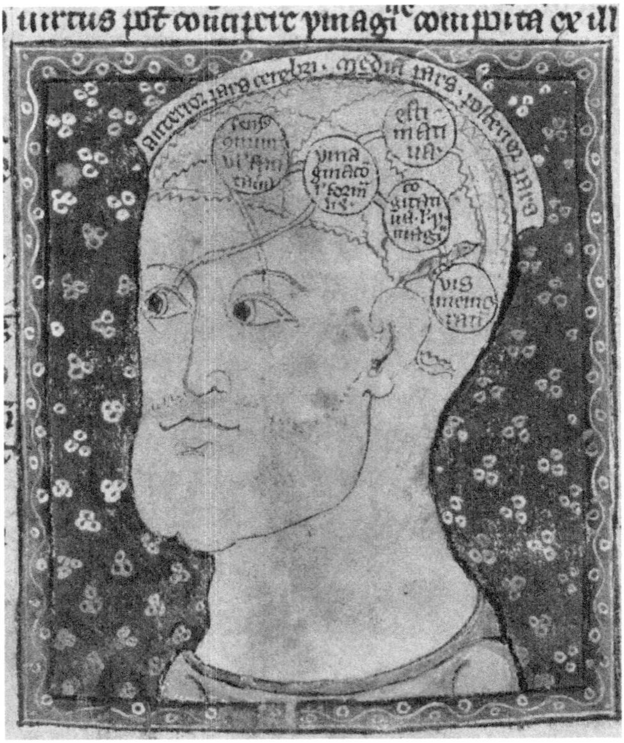

Figure 4
Cambridge,
University
Library, Nn 3.2,
fol. 36v. Public
Domain.

Figure 5 Hildegard of Bingen, *Scivias*; Rupertsberg Codex, f 1v.
Public Domain.

Figure 6 Hildegard of Bingen, *Liber divinorum*, Biblioteca Statale Lucca, MS 1942, fol. 28v. Via © World Digital Library Collection, via Library of Congress. Public Domain.

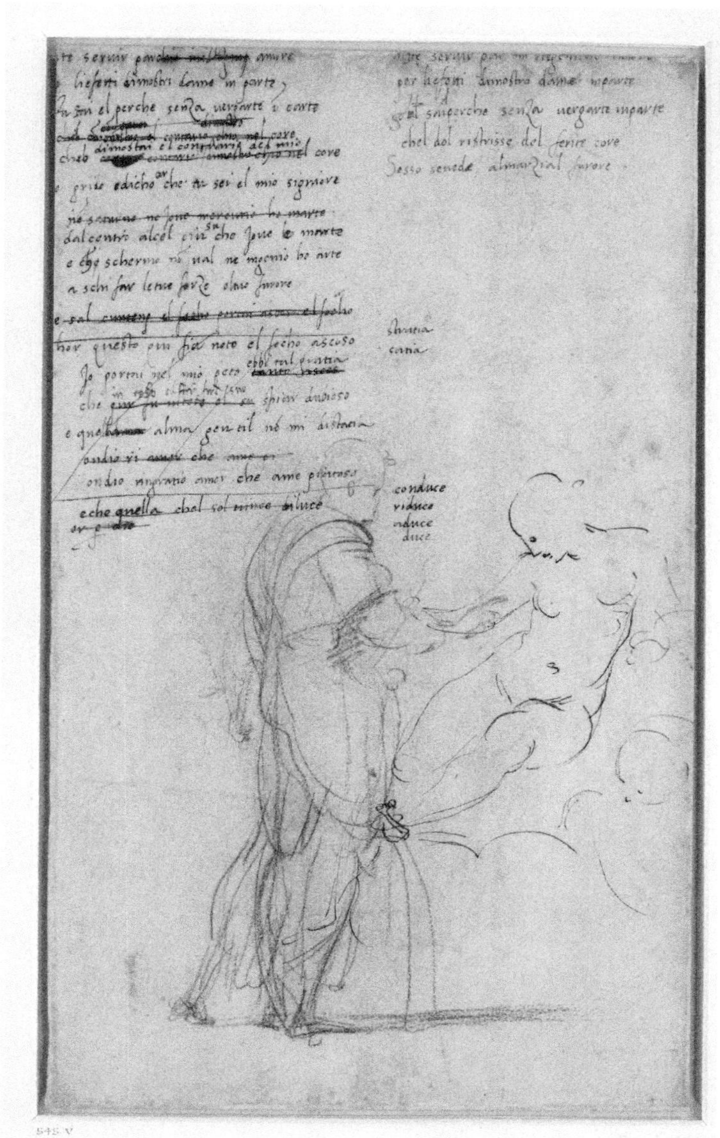

Figure 7 Raphael, drawings. Ashmolean Museum, WA1846.186 "Verso: Figure Studies and a draft sonnet." © Ashmolean Museum, University of Oxford.

Figure 8 Oxford, Bodleian Library, MS Auct. F 6.5, fol. 1v.
Reproduced by permission.

Figure 9 Munich, Staatsbibliothek, Clm 15825, fol. 1v. CC PDM 1.0.

Figure 10 Utrecht, University Library, Ms.1335 (2 C 15 fol), fol. 178bis v. Open Access via Utrecht University Repository.

Figure 11 Cambridge, Fitzwilliam Museum, "The Macclesfield Psalter," MS 1-2005, fol. 115v. Photograph © The Fitzwilliam Museum, University of Cambridge, reproduced by permission.

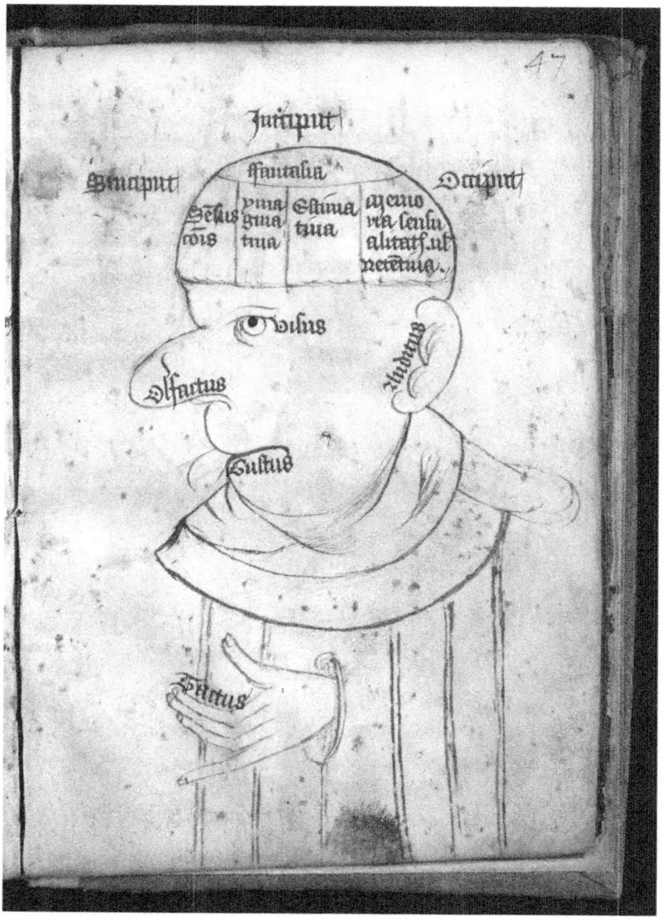

Figure 12 Cambridge, Trinity College Library, MS O.7.16, fol. 47. © The Master and Fellows of Trinity College, Cambridge. Reproduced by permission.

Figure 13 Paris, Bibliothèque nationale de France, MS Fr 19093 fol. 18v. Public Domain.

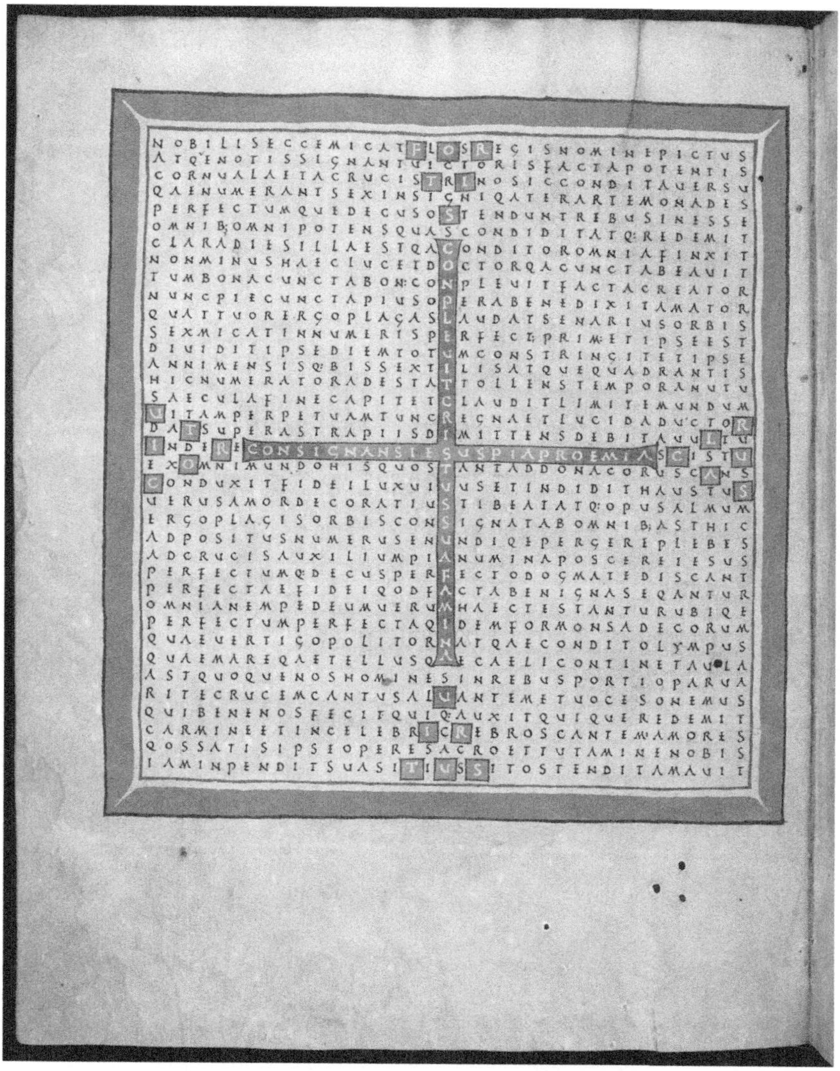

Figure 14 Hrabanus Maurus cross, Cambridge, Trinity College, MS B.16.3, fol. 25v. © The Master and Fellows of Trinity College, Cambridge. Reproduced by permission.

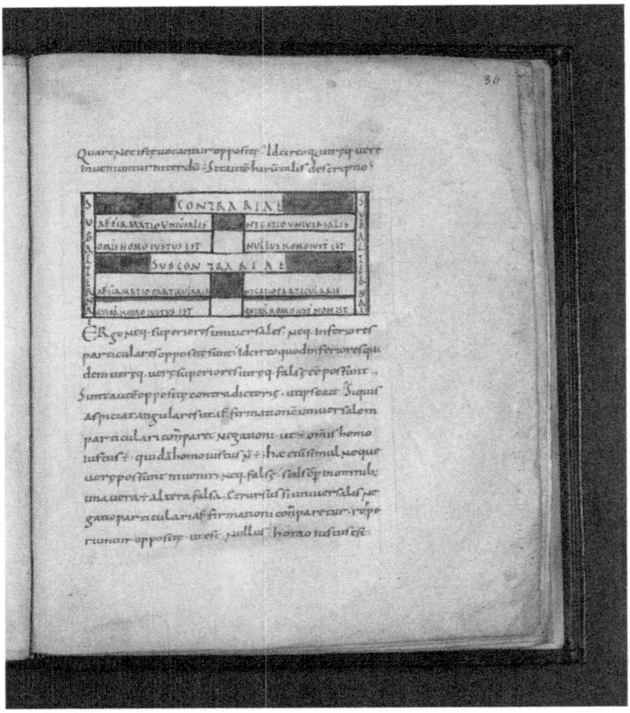

Figure 15A Philadelphia, University of Pennsylvania Schoenberg Library, MS LJS 101, fol. 36.

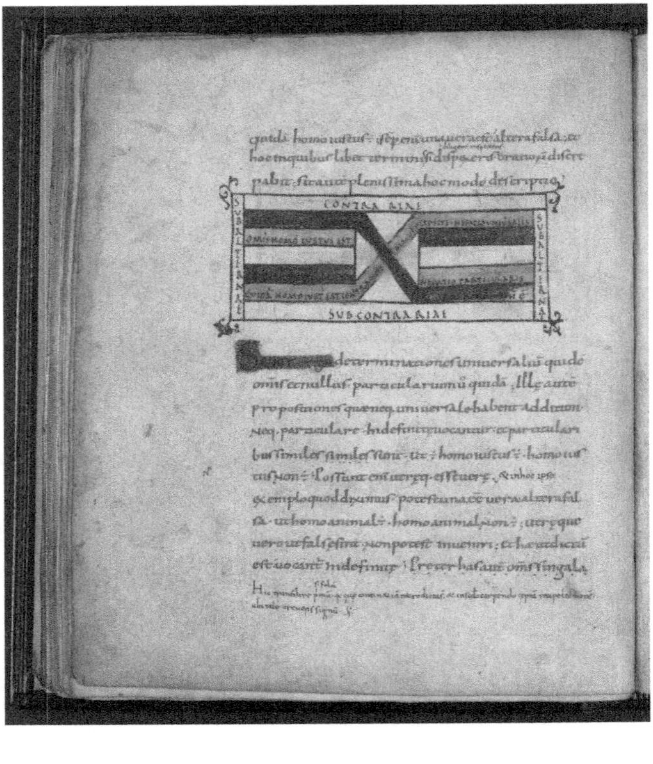

Figure 15B Philadelphia, University of Pennsylvania Schoenberg Library, MS LJS 101, fol. 36v.

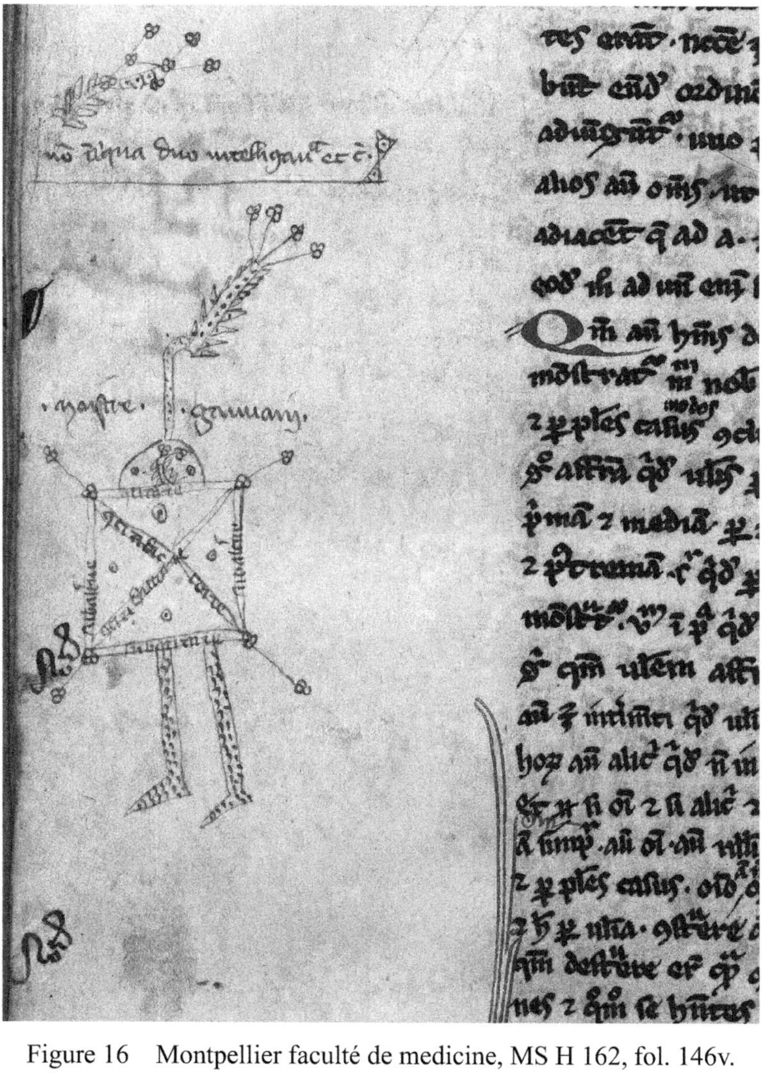

Figure 16 Montpellier faculté de medicine, MS H 162, fol. 146v. Public Domain.

Figure 17 Munich, Staatsbibliotek, Clm 14731, fol. 80. Public Domain.

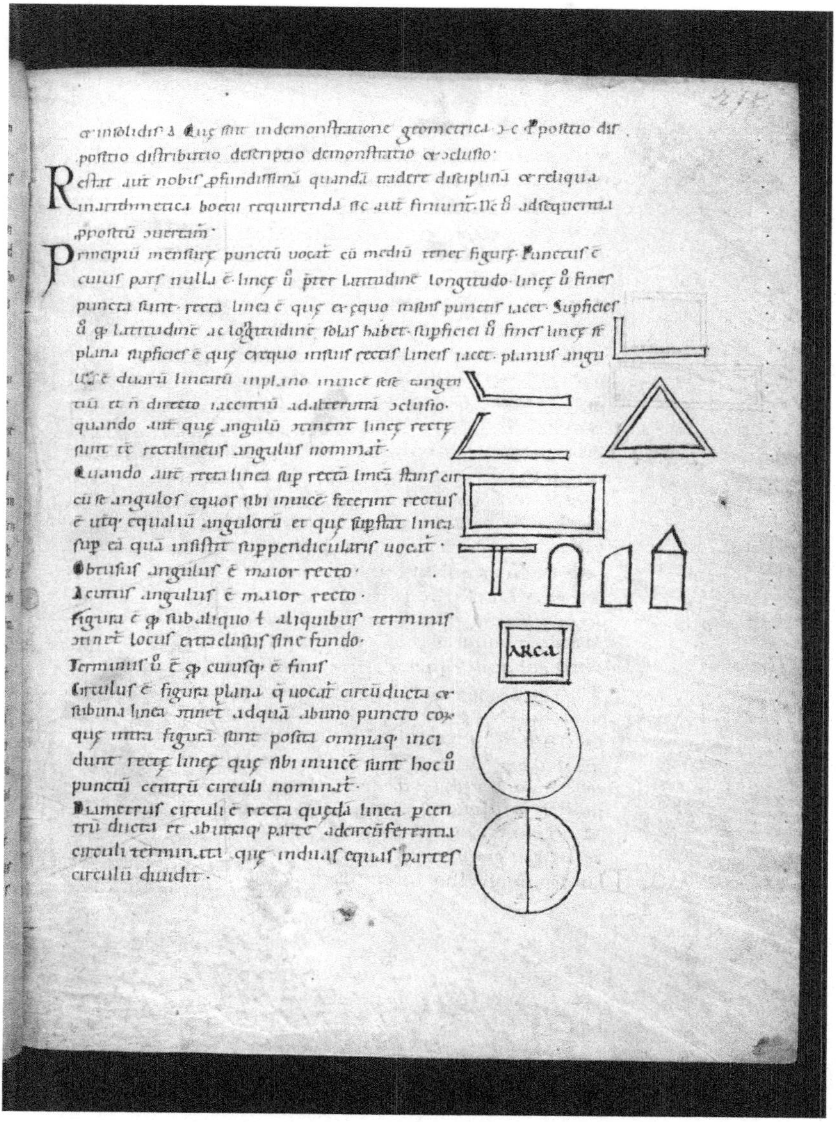

Figure 18 St. Gallen, Stiftsbibliothek, Cod. Sang. 830, p. 295. Reproduced by permission.

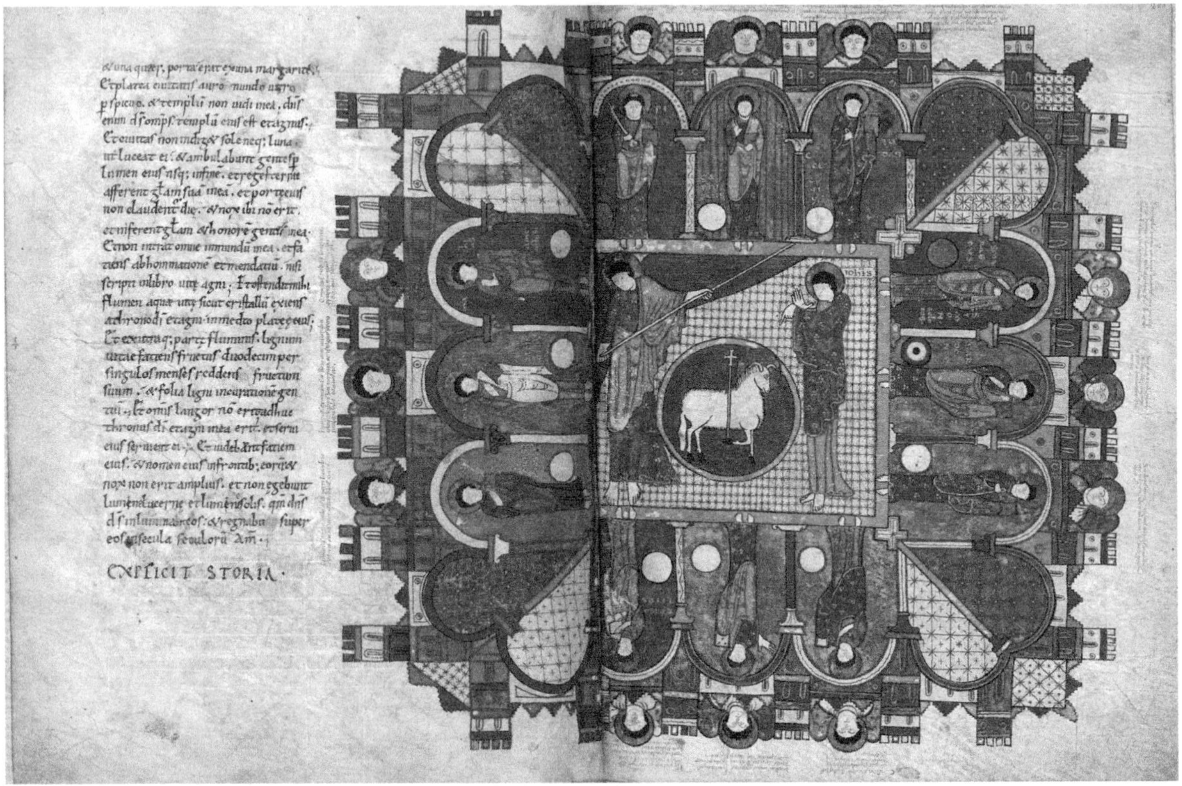

Figure 19 Paris, Bibliothèque nationale de France MS lat. 8878, fols. 207v–208. Public Domain.

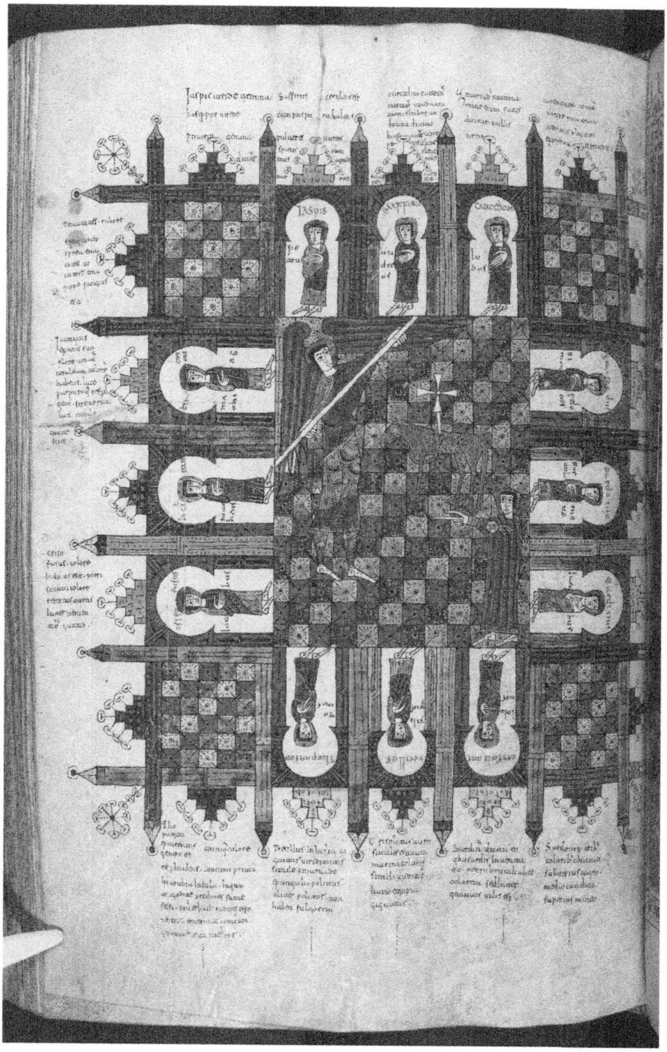

Figure 20 London, British Library Add. MS 11695, fol. 208v. From the British Library Archive, reproduced by permission.

Figure 21 London, British Library, MS Burney 275, fol. 293. From the British Library Archive, reproduced by permission.

Figure 22 Rushton, Northants. "The Triangular Lodge" (photo: M. Carruthers).

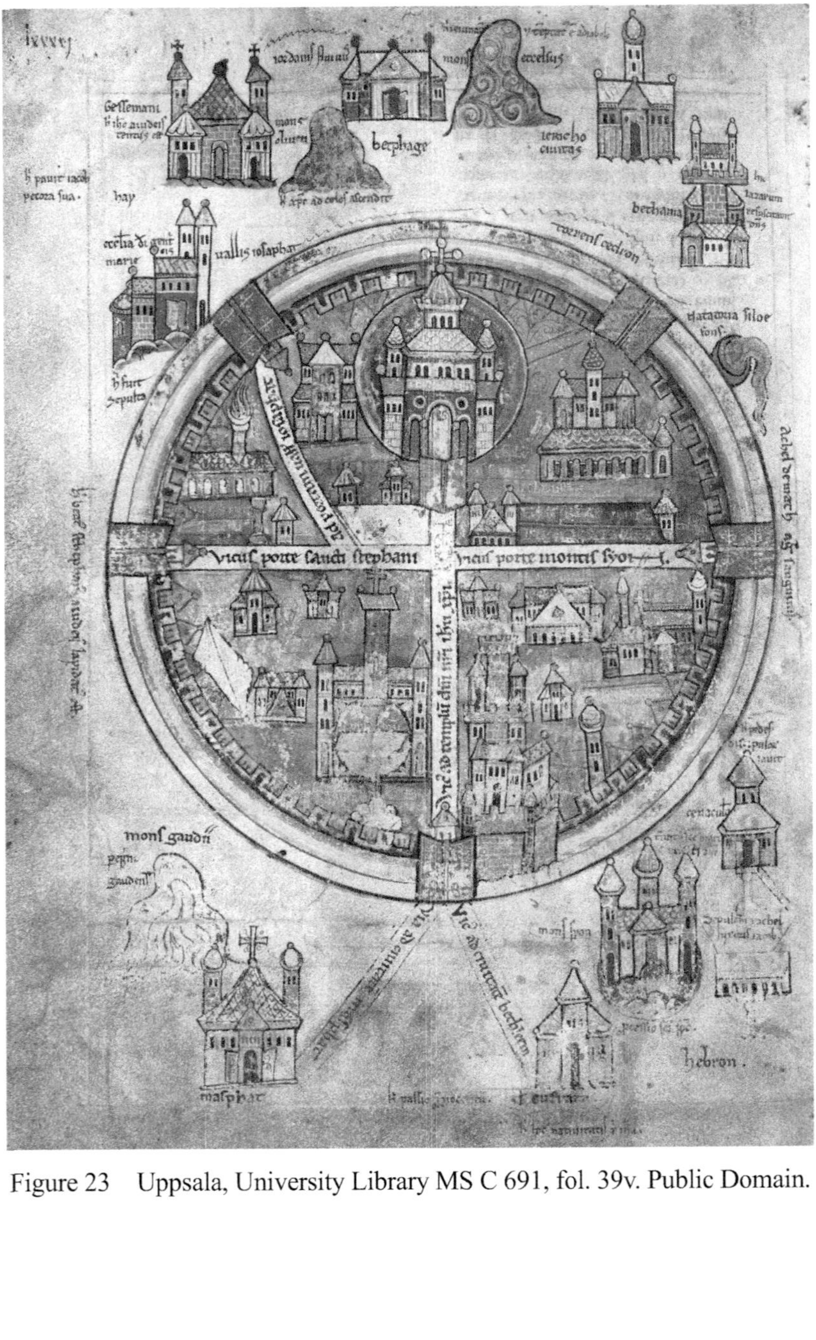

Figure 23 Uppsala, University Library MS C 691, fol. 39v. Public Domain.

Figure 24 “Typus Gra*m*matice”: G. Reich, *Margarita philosophica* (Strasbourg, 1504). Via Google Books, image in the Public Domain.

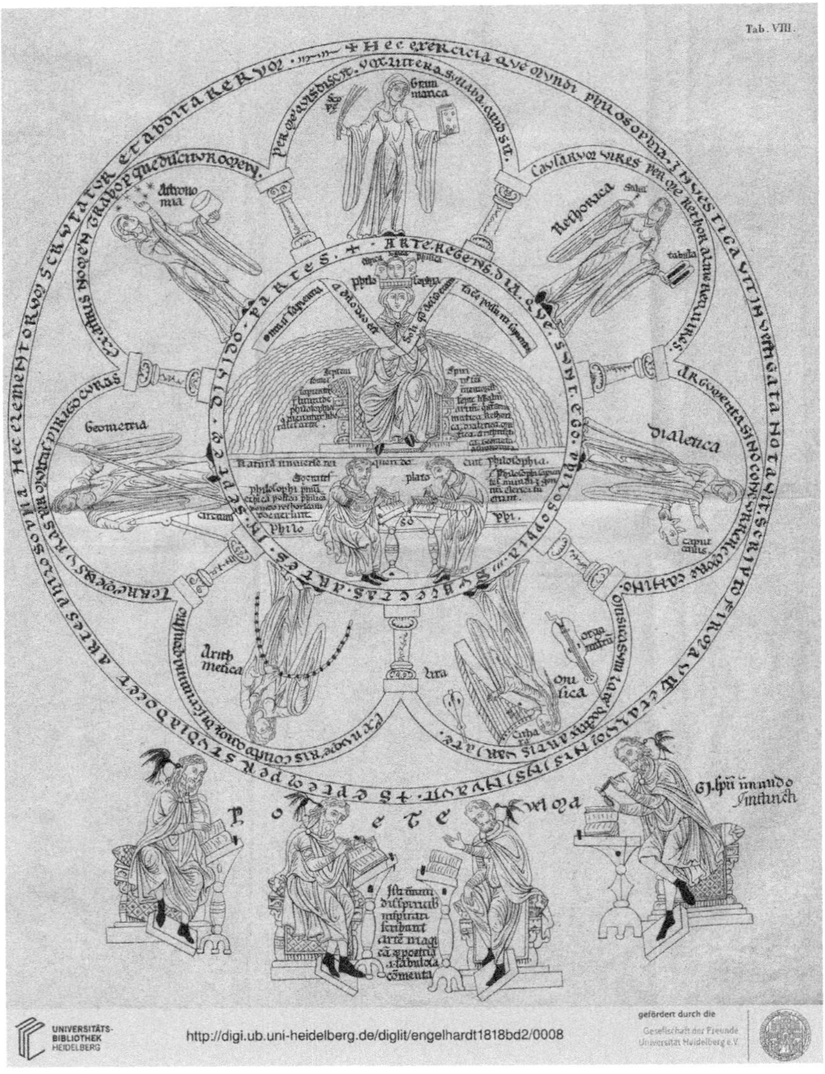

Figure 25 Herrad of Landesberg/Hohenbourg, "Hortus deliciarum," tafel 8. From a drawing of the original MS by Christian Moritz Engelhardt (1818). Special Collections, Heidelberg University Library, Public Domain.

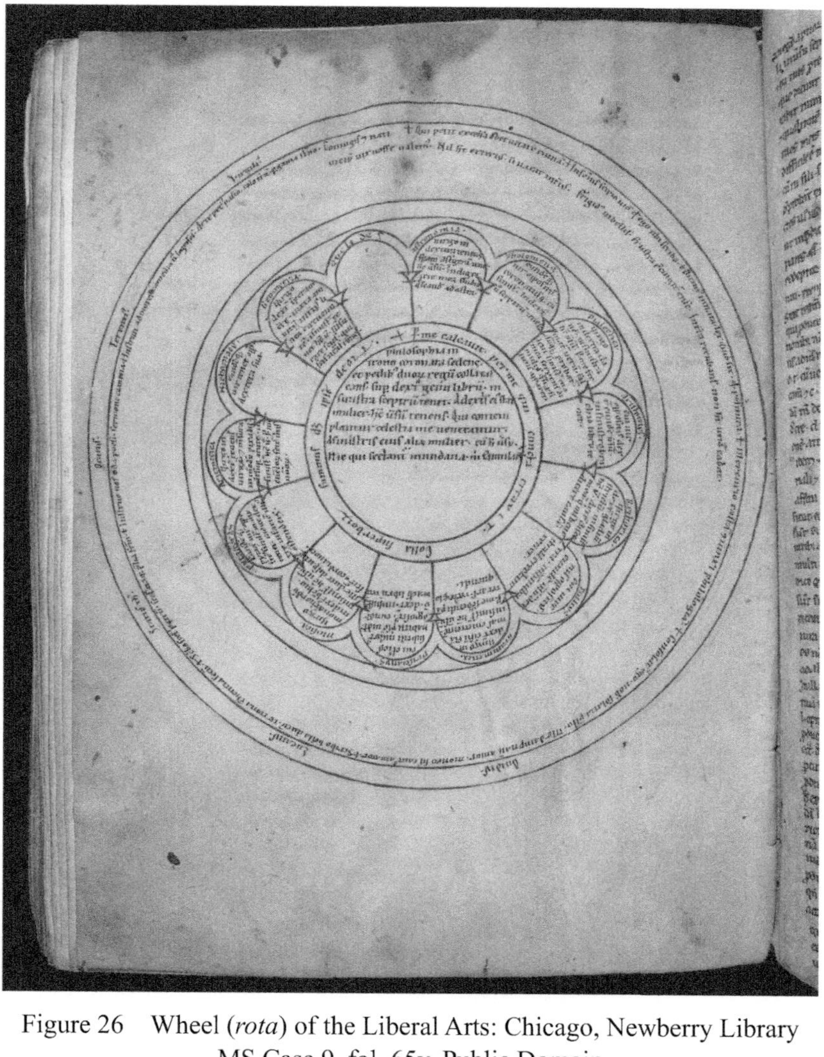

Figure 26 Wheel (*rota*) of the Liberal Arts: Chicago, Newberry Library MS Case 9, fol. 65v. Public Domain

Introduction: Creative Geometries

> Really, universally, relations stop nowhere, and the exquisite problem of the artist is eternally but to draw, by a geometry of his own, the circle within which they shall happily appear to do so.
>
> Henry James, preface to *Roderick Hudson*

This book is about beginnings, both of human crafting and of cosmic Creation. Because of that, it is also necessarily about the processes of forming, designing, and crafting thinking – the complex of procedures involved in what is called *poeisis* in Greek, *faciendi* and *poesis* in Latin, and both "fashioning"and "making" in English. For Cicero and for Horace, and the many centuries of authors who followed them, *ars poetica* was not confined to fashioning metrical verses, but is an extensive title, including all of what we now know as "language arts" (and have come to rigidly separate from one another, as well as sharply from all the "mathematical" and "scientific" arts of the *quadrivium*). Aristotle remarks at the start of his *Metaphysics* that philosophers must also be myth-lovers (*philomuthos*) "since myths are composed of wonders" and therefore will set us to wondering about the world around us, a remark that certainly includes Plato as well as Pythagoras, both of whom were myth-lovers.[1] In the Latin school manuals used for centuries to teach rhetoric, procedures for fashioning language are called *dispositio*, in English "Organization"; most commonly in medieval Latin they are collectively called *ordo* or *ordinatio*, and humans have invented and used their own orderings for their understandings of the ordered universe in which they find themselves.[2]

1 Aristotle, *Metaphysics*, I. ii (982b); translated Tredennick, vol. 1, p. 12.

2 Hicks 2017, p. 33, points out that there is a radial pun (for that is what an inventive rhetorical *etymologia* is), which was recognized in late Latin that translates the word-play of Greek κόσμος ("universe, order") and διακόσμησις ("ordering, arrangement") into an equivalent Latin play of *ornata* ("arranged, ornamented") and *ordinatio* ("ordering, regulating"). This Latin verbal trope occurs in Apuleius, often in Augustine and later writers on cosmology like Boethius, thus bringing

God speaks ("dixit Deus") in the very moment that God also "fashions" ("Deus *fecit*"), as Genesis 1 recounts on each of the six days of divine creating. God creates the cosmos through Sounds that also make Lights – harmonious sounds forming ordered rational, even chordal patterns, a fundamentally "Pythagorean" idea.[3] But, as Genesis 2 makes clear, humans *name* the creatures of Earth (Gen. 2: 19–20). God brings all the living beasts and birds to Adam out of a sort of divine curiosity, because apparently God wishes to see what Adam will come up with by way of his human speech-labels to differentiate each kind of animal *as it appears to him*.[4] Adam's names have nothing to do with whether these creatures actually live or not, for God has given them life. Adam just measures them out into his own labelled, nominative categories. Commenting on this account in Genesis,[5] St. Augustine notes that God did not bring any fish to Adam for naming, and wonders why this would be. He explains that the fish (and the seas, for that matter) were unknown to Adam in Eden, which had rivers but no oceans. So Adam and his descendents would have to give names to the fish and to other animals and plants unknown in Eden, which they encountered after humans had been expelled into the greater world. As voluble descendents

humanly inventive linguistic play (Rhetoric) into reflective harmony with the principle of cosmic ordering as understood through *Mathematica*. In such ways all the Liberal Arts were said to "participate" in the divinely harmonious motions of Creation. Hicks's study of medieval cosmology is an essential complement to much of my own. *Cosmos* also incorporates the values of order and ornament incorporated in "beauty"; even contemporary English "cosmetic" retains a distant though debased relationship to *cosmos*. Palazzo 2024 explores the idea of κόσμος in terms of embroidery and weaving.

3 A succinct, readable summary of ancient Pythagorean ideas of mathematically expressed cosmic harmony is in the introduction by H. Tredennick to his LCL volume of Aristotle's *Metaphysics*, pp. xiv–xix. The basic ancient Greek idea was of "the universe as a living organism" (p. xv), composed of parts that are not singular beings but each a "participant" in a Whole cosmic body, expressed through measurable *ratio* and rhythm as 'harmonies,' which extended to human cultures and societies also. I will return to this basic idea often in the chapters that follow. See also Cornford 2000 (1937), his masterful "running commentary" on Plato's *Timaeus*. See also Lapidge 1980.

4 Gen. 2: 19: "Formatis igitur Dominus Deus de humo cunctis animantibus terræ, et universis volatilibus cæli, adduxit ea ad Adam, ut videret quid vocaret ea: omne enim quod vocavit Adam animæ viventis, ipsum est nomen ejus." [Now the Lord God had formed out of the ground all the wild animals and all the birds in the sky. He brought them to the man to see what he would name them; and whatever the man called each living creature, that was its name. So the man gave names to all the livestock, the birds in the sky and all the wild animals.]

5 Augustine, *De Genesi ad littera*, IX, cap. 12, 20.

of the first humans, they continued to name them in their own different post-Edenic languages. The tower of Babel incident (Gen. 11: 1–9) makes it clear, Augustine says, that God wishes humans to fashion their own languages in accord with their fruitful multitudes; the same living animal and plant can have many different names in various human languages. But their various human names do not give them life. Human naming is an activity essential only for human reasoning, human-devised *ratiocinatio* as it is called in logic and rhetoric texts. It is the nature of our minds to categorize and thereby rationalize our Earth as we can. Human reasoning is not and cannot be coterminous with creatures, as God has created each of them, but at its best it might in its limited and partial manner "reflect" a few elements of that divine order. One such reflecting mirror that human reasoning has fashioned for itself is Geometry, *Geo-metria*, the human science of Earthly measuring, both *of* the Earth and *by* Earthly means *in* human perspectives from the Earth.

Geometry was understood to be a mathematical subject fundamental to all manner of verbal crafting as well as to numerical measuring and to those ethical, judicial, and medical values emphasizing moderation and balance that we think of as "classical." In the words of the book of Wisdom (11: 21), when creating, God also arranged and disposed, *disposuisti*, all things in rational terms with measure (dimension) and number and mass/weight [*omnia in mensura et numero et pondere disposuisti*], a maxim much referenced throughout medieval centuries. Human thinkers, seeking to understand Nature through their own languages, including their understandings of numbers, were considered to create similarly, though in their lesser, limited human fashioning. Pre-modern education simply did not divide literacy so sharply from numeracy as our post-Enlightenment elementary education curriculum does, even to the point of insisting that the two involve entirely different mental processes. The classic curriculum of the Liberal Arts made no such assumption.[6] Since I

6 Albertson 2014, pp. 40–59, focuses on this point in his discussion of the Neopythagorean writers Philo Judaeus (d. 40 CE) and Nicomachus (d. 160 CE), both of whom worked in Alexandria (a crucial nurturer and blender of late antique Greco-Roman-Judeo-Christian thought, and a major conduit still even centuries later for the Indo-Persian-Arabic scholarship that so enriched the already wondrous late antique philosophical stew). Philo is credited as the earliest proponent of divine Logos as "the agent that distributes God's transcendent equality and monadic power to the enumerated order of creation" (p. 47), thus blending biblical accounts centered about the mathematical harmonies described in the books of Moses as having been commanded by God, with Greco-Roman mathematical Idealism, the belief that divine Arithmos is cosmically expressed in harmonies whose numbers are rationally comprehensible by human intellect. Nicomachus "not only invented the comprehensive number theory that would in time become the medieval quadrivium, but also

will be discussing the verbal nature of geometrical reasoning throughout this book, readers must always keep this in mind. God creates the natural world through the mediating powers of both *Logos* and *Arithmos*, and to divide the two is profoundly to misunderstand them both.

The verbal arts of the *trivium* act synergistically with the mathematical arts of the *quadrivium*. Indeed the language arts as a whole were also at this time called the *logica* ("logics," from *Logos*). Music, a mathematical subject (*Arithmos*), included all manner of articulated harmony and rhythm, including poetry and metrical prose (*cursus*).[7] Astronomy was taught in part by means of fables recounted in carefully metered verses (the better to learn and recall), chiefly Aratus's *Phaenomena*. In pre-modern western cultures, mathematical reasoning was expressed mainly in ordinary sentences until the tenth century (Augustine provides some excellent examples, as we will see in Chapter 3, but he is not alone). This is one of the most difficult things for readers now to fully appreciate. Quintilian counsels at the beginning of his *Fundamentals of Oratory* (I.x.35ff) that an orator should have mastered four subject matters essential to the craft: grammar, music, acting – and geometry. The intimate relationship between mensuration and all manner of human crafting is a continuing theme of this book.

---

juxtaposed mathematical mediation with Logos-mediation for the first time" (p. 40). Albertson offers a readable history of these difficult theological cross-currents, particularly as expressed through the curriculum of the late antique Liberal Arts in a manner considered to blend verbal and mathematical methods of analysis (*trivium* and *quadrivium*). Several scholars have attended to the matter of bridging the wide Modern gap that has appeared between the supposed "two cultures": two that try to address this as an issue involving rhetorical Style as well as analytical Grammar and Logic are T. Williams 2016 and Rotman 2000. Augustine's dialogue *De ordine* (Bk. 2, esp. 11–15) presents a fundamental discussion of how sound (received in the ears) is understood (measured) first by applying units of relative temporal length and emphasis (Grammar), then, building on these elements, as Music and as Geometry. These sounds also can be communicated systematically through readable visual signs or *litterae* ("letters"). This late antique sequence of the liberal arts, which begins with sound (*sonus*) as grammar and logic, music, and geometry, is also articulated in the elementary grammars of Donatus, Priscian, and of Boethius's treatises both "on arithmetic" and "on music." These, together with Aristotle-based treatises of the logical *Ysagogue* and *Organon*, made up the elementary curriculum of the monastic schools; see Riché 1978, 1989.

7 Bower's translation (1989) of Boethius's *Fundamentals of Music* brings out the grammatical as well as numerical nature of music particularly well; see also Bower 2001, and Leach 2007. Medieval vernacular poets as well as clerical teachers took the musical/numerical nature of poetry for granted, a connection evident in the fact that the works of *poetae* were referenced throughout the curriculum.

The linguistic art of Invention is fundamentally a *techne* for crafting and making (*ars inveniendi*) and not only a discretionary art (*ars judicandi*) of defining and categorizing a thing already finished. Beginning also implies some goal, its *scopus* ("target") that the artisan desires to achieve, and also an informing shape chosen for getting there, however indirectly (and even if, as in some modern art, the goal is to appear aimless, for even chaos has an implicit grammar – otherwise we couldn't recognize it as "chaotic"). The art of Invention is concerned first with *origines* as both "origins" and "sources," the *auctores* or "wells" of a creative stream. There is no stream without its sources, yet a source does not exist all by itself – it requires its own flowing streams in order to be considered a source at all. Nineteenth-century Romantic uses of the word "original" employ a different, almost a reverse, model of this pre-modern one, products as they are of the Great Renaming that constitutes such a large part of western Romanticism and Modernism. One now reads claims that something is "uniquely original," as though "unique" – which means "singular, one of a kind" – and "original" were of the same semantic family.[8] But for medieval thinkers a proper synonym of *origin* is not *unicus* but *fons, fontis*, a "spring, fountain" for many others, not a singular one-off.

Cassiodorus (d. 585) parsed the starting point for all our coherent knowing and inventing as the judging method known from Aristotle as *Topica*, a matter of identifying our "origines sensuum et fontes dictionum," the sources of our sense-based knowing and the fountain-heads of the matters of which we speak.[9] Medieval authors took seriously the root meaning of the verb *invenire*, "to start on" a process and also a journey. As is true of all journeys, their routes have some implicit *ratio* (which means "plan" and "pattern" as well as

8 Latin *origo, -inis* is derived from the same root as *oriens*, the daily eastern "dawnings" of the sun in all the great varieties we are so familiar with.

9 A clear exposition of this ancient model in the ethical and cognitive thought of St. Augustine is Pelikan 1986. As he elegantly pointed out, the Socratic command "Know thyself" (always emphasized in the many forms of western monastic meditation) starts with imitation (as psychology now understands it), learning first from others and thereby extending one's experience and judgment outwards by analogy with others; see also the account of Augustine's "Christian socratism" in Gilson 1960. The notion of all generative movement as extending outward in time as well as space is basic not only to ideas about the human soul and self but also to cosmology, derived in part from *Timaeus*. See Sorabji 1983 and Colish 1990, both of which stress the Stoic elements in these ethico-physical ideas, not just looking to Platonist originals. Equally fundamental in hexaemeral accounts of the cosmos (such as Augustine's) is the Aristotelian belief that its essential nature is continual movement and change; it is thus both temporal and temporary with respect to unchanging Eternity. A clear account of this matter, by a physicist who is also an eloquent exponent of his subject, is Rovelli 2018.

"reason"), an order that will shine through as it is fully realized when one can finally understand it whole. In the terminology of classical rhetoric, the beginning of a composition is its *exordium*, from *ex* + *ordo*. As a craft term, Latin *exordium* refers also to the warp threads laid out first on a loom when cloth is woven, an analogy that augments the very meaning of the word "text" itself, which is from *textus*, "woven." In the verbal crafts of meditative thought and language weaving, those first threads are vital, for they are the only true begetters of the fully completed tapestry as we know it.

Medieval invention focuses on generation, not on end product, using as its fundamental model the process of embryos forming into babies. It emphasizes initiation over ending (though of course the process presupposes that some goal, some aim and *scopus* as both "target" and "scope of work," has also been planned out in the artist-weaver's mind, as the cosmos was in God's mind "in the beginning." Geoffrey of Vinsauf called it the "Cadiz" or final port, the desired conclusion, of the textual journey that he uses as his model for the crafting of poetry, with the poet as both the map-maker and fellow traveler together with the audience.[10] Medieval *origines*, both divine and natural, are neither solitary nor one of a kind but progenitors of large families of kin born variously from their initiating source. This is a model of natural creation that emphasizes movement, change, and birth, not "timeless" rules. Natural and human creation can never be static. Indeed Latin *principium* means "start"; "principles" are not transcendently universal laws that things of the same kind invariably must follow but the "initiators" (though within their implied boundaries) of much fruitful, various work yet to be done. It is worth keeping in mind that, in medieval theology, the elemental creating motion that mediates and translates the divine plan is curving, not straight. It bends and turns circularly in spirals, as does the universe it has "started off" so that all the embryonic motions of the cosmos can grow out from it. All subsequent cosmic life, including that of angels, and all human understanding depends on the essential movement and moment "in the beginning."

In his greatly influential *Commentary* on the first two books of Plato's theological speculation about natural creation, *Timaeus* (and using his own translation of these into Latin), Calcidius (around 500 CE) described

[10] See note 14, below. I discussed in Carruthers 2010 the much-emphasized medieval concept of a composition's *ductus*, how a work should be planned like a journey (allowing for many pauses and byways) so as to guide its audiences through itself from its start to destination without its travelers just wandering off or dropping out, and likely to become hopelessly lost (the root meaning of *error* is "wander").

how human understanding is shaped by its very nature in a fashion similar to physical creation. In Calcidius's Latin, God is called *opifex deus*, the divine *artisan*. This God desires to make things (being the blacksmith as well as judge of his handiworks). As God created the cosmos he did so in ways that would be comprehensible to its creatures, to the extent they were capable of such understanding. "Suitability" – *decorum* and *ornamentum* – is simultaneously a cardinal system of aesthetic and ethical values in the physical universe as it is also in crafting, both by humans and by animals such as bees and birds. Plato's living World Soul (*anima mundi*) animates *hyle* (matter) and generates all natural forms "out of the spins and curves of its thought and remembering."[11] But, like all changing things, the cosmic revolutions are less than perfect. There could be no natural creation otherwise, for movement is born from movement, and Eternal Being, perfectly complete within itself, has neither motion nor time – unless and until Being itself might desire to start off its creating sequences. Natural creation moves in spiraling gyres because its curving movements make less than a perfect return to their beginning point. Plato explains the errant motion of planets as owing to "their turning round in a spiral and, as it were, a winding acanthus coil [=Latin *volumen*]."[12] *Volumen* is also Latin for a scroll, a book that must be unrolled to be read.

These are fundamental ideas in Augustine as well. Human thinking *revolves* and turns: the first word in Augustine's self-dialogue *Soliloquies* is *volventi*, "turning about": "Volventi mihi multa ac varia mecum diu ac per multos dies sedulo quaerenti memetipsum ac bonum meum [turning over for a long time to myself and with myself many and various things, and through many days seeking assiduously my very self and my good]."[13] This is the principal movement of Augustine's Nature too, as the One Divine Mind both speaks (*vox*) and radiates (*lux*) into life the cosmos in all its numerous, manifold varieties, via the divine Word, its own life-creating Plan. And such circling, according to Geoffrey of Vinsauf, is the initiating motion of creating poetry too: "Let the mind's interior compass first circle the whole extent of the material. Let a

[11] Calcidius *Comm.* I.101 (on *Timaeus* 36e): "cogitationis recordationisque gyris et anfractibus [=bendings]" (ed. McGee, 290–1). McGee's introduction is an excellent guide to the many unsettled problems of authorship, date, and sources of the Commentary, plus an overview of the manuscripts. Further references are to this edition.

[12] Calcidius, *Comm.* I.116, on *Timaeus* 39a (ed. McGee 310–11): "illa conversio [=turning] vertens in spiram et velut sinuosum acanthi volumen." The animating "breath of life," the *spiraculum* of Gen. 2: 7, moves in a spiral; see Palazzo 2020.

[13] Augustine, *Soliloquiae* (PL 32.869).

definite order chart in advance at what point the pen will take up its course, or where it will fix its Cadiz."[14]

Lines and circles differ in shape but not in kind from one another, for they both result from *a terminated movement that has left a trace that we are now able to measure*. That is the essential concept, the initiating idea by which all natural forms are imagined and defined in terms suited to the natural boundaries of human reasoning. The divine craftsman, *opifex deus*, having carefully pre-visioned and analyzed the things he desires to make, takes his tools and with those tools draws the in-forming limits (the *termini*) of the various shapes he has envisioned and will proceed to construct. The two essential tools an *opifex* requires for measuring are a straight-edge and a compass, shown often in medieval illustrations of creation, both when divine and when human (see Figures 1 and 4). This Platonist idea of Nature as *technon*, an *artisan*-creator realizing, though imperfectly, a divinely initiated *ratio* articulated in Word and Number, is essential, underscored by all the parallels made in these cultures to human crafting. Calcidius speaks of God not only as *opifex* but also as *architectus*, "architect." And in ways that link the artisan model strongly to all the language arts, to Rhetoric as well as Dialectic and Grammar, Calcidius emphasizes the innate "desire," as he calls it, of material (*hyle/silva*) to be shaped and *ornata*, both ordered and ornamented.[15] In his discussions of the Genesis account, Augustine frequently expresses his understanding of the divine order in complexly ordered, elegantly crafted sentences (I discuss this further in Chapter 3). Calcidius calls the initiating cosmic dilation (as described in Gen. 1: 1–2) an *exordium*, the preface to a well-planned (though still incompletely executed) oration. What binds language and mathematics in this model is their shared principle of syntax, Greek *συνταξις*, ordering and arrangement. Though these orderings are not the same, they grow from

[14] Geoffrey of Vinsauf, *Poetria nova*, I, 55–58: "Circinus interior mentis praecircinet omne/ Materia spatium. Certes praelimitet ordo/ Unde praearripiat cursam stylus, aut iti Cades/ Figat"; trans. Nims, p. 20. Only as one orders one's materials do its in-forming lines move within this initial measurement to form the *viae* and *itineraria*, paths with itineraries, of the completed work.

[15] This idea, that matter *desires* to be formed, is prominent also in Bernardus Silvestris's *Cosmographia* (ca. 1147) and is a recurrent theme in twelfth- and thirteenth-century writing. *Hyle/Silva* is a female principle in these accounts. Dante expresses this principle (and assigns it to the *tre donne* who initiate his vision) in his emphasis on desire and love as the attracting and thus animating energy of the entire cosmos. Like many other iterations of the idea, it is prominent in Augustine's hexaemeral thought also. Augustine stresses that Nature's original "goodness" (as characterized in *Gen.* 1) was fully coherent, physically powerful, ethically wise, and aesthetically pleasing.

a shared principle, as Calcidius and Plato emphasized. And this is why, in a study of the verbal arts of composing, some attention needs to be given to ideas about physical Invention also, and to Arithmos as well as Logos.[16]

Geometry was considered to be a rational instrument devised by rational humans in order to understand the world they live in. God first planned, implicated, and then expressed the world from his Eternal mind, which by definition is far beyond any human ability to comprehend. So human reasoning has abstracted qualities and quantities of orderliness and comeliness from what God created in ways (the methods of grammar and mathematics) compatible within the bounds of human reasoning and intellect. Human knowing is only a mirror image of Divinity, and for that reason it is always partial, limited, and also reversed – seeing the divine only through puzzles comprehended darkly as in a mirror. All the geometric figures, called *geometricalia* in Latin, are themselves only mental abstractions drawn by human minds from the created forms existing in nature.[17] And yet, by their very nature, human beings *desire* to know their creation and Creator.[18] God created all natural beings in their particular forms: needing to understand them, humans invented Mathematica (as they also invented Grammatica), the chief tools of their imperfect, limited, but naturally wondering and curious minds. A much-quoted maxim deriving from Aristotle's *Metaphysics* holds that human philosophizing begins in acts of wondering and puzzling; companion to it is one deriving from his treatise on *On the Soul*, that humans cannot reason without some visualizable structure – geometric figures, number sequences, and written *litterae* representing sounds provided obvious ones – with which to articulate their thoughts.[19]

16 I learned much about both Arithmos and Logos as theological principles especially from Albertson 2014, and McGinn 2002. I first began thinking about these matters when reading Stock 1972 and Wetherbee 1972, and later Dronke 2008.

17 In this way among others, late antique and medieval hexaemeral theology differed sharply from *Timaeus* (and from sixteenth- and seventeenth-century Platonism), insisting, with Cassiodorus, that mathematical objects are humanly made rational abstractions from created Nature and not in themselves Real and unchanging Ideas. See Rovelli and Albertson; also works cited in n. 11 above. I discuss this matter in greater detail in Chapters 3 and 4, below.

18 This is the crux of Augustine's account of the Fall as the result of human failure to grasp the initial and initiating role of desire in all human ethical judgment (*arbitrium*) and choice (*voluntas*), human "knowledge of good and evil"; see *De Gen. ad litt.* XI. 4–10 esp. By not understanding this, both Eve and Adam in their human vanity and willfulness were susceptible to errors of judgment and consequent choice.

19 See *Metaphysics* I.ii (982b) and *De anima* III.iii (427a–b). Versions of these maxims well suited the Genesis story of divine creation, as Adam and Eve's human

Cassiodorus defined *mathesis* as "the body of knowledge (Latin *doctrina*) that considers abstract quantity; and a quantity is called abstract which, having been separated from its material and all other accidentals by the intellect, we derive by reason only."[20] This definition of mathematics (*mathesis*), quoted by Isidore of Seville among many others, was standard throughout the Middle Ages.[21]

Human minds understand Nature through memory and rational measurement, which together allow us to trace out our world in the *vestigia*, the "tracks" of moving lines. It is a common idea about the human mind, expressed in etymological terms by many medieval writers, that Latin *mens, mentis* (mind) is part of a conceptual family which includes *memoria*, "memory," and also *mensura*, "measure." And our reasoning intellect's boundaries lie precisely within the abilities of these two mental instruments, for we humans cannot measure things that are always changing in a temporo-spatial blur from one instant to the next. For our minds to measure any of the things we perceive in space and time – and thus enable ourselves to comprehend time's traces – we

desires and judgments, good and poor, all start from acts of wondering. Basic practices of religious contemplation dependent on geometric structures were in place well before Christianity.

20 Cassiodorus, *Inst.* II praef. 4 (ed. Mynors, 92): "mathematica vero est scientia quae abstractam considerat quantitatem; abstracta enim quantitas dicitur, quam intellectu a materia separantes vel ab aliis accidentibus, sola ratiocinatione tractamus." Cassiodorus repeats this definition at several points in Book II. Isidore of Seville quotes it as his definition of *mathematica*, *Etymol.* III praef. It should be kept in mind that *mathesis/mathematica* includes all the subjects of the quadrivium: arithmetic and geometry, music and astronomy, just as *grammatica* (also called *logica*) includes the three subjects of grammar, dialectic, and rhetoric. Pre-modern terms cannot be held to modern standards of extreme categorization.

21 In *De Gen. ad Litt.* IX, xii.20, commenting on why Adam, rather than God directly, names the animals presented to him in Eden (Gen. 2: 19), Augustine makes an early version of this logical point. To name something is to label it, an act necessary in human reasoning and judging, discretion and discernment. Adam's names for the animals must also have been in whatever language he spoke in Eden. But that language is now unknown and unknowable, since after the Tower of Babel, human languages are all various and mutually incomprehensible. The animals' names are thus both arbitrary and variable, limited in the manner of all human reasoning. Adam also names woman (*mulier*) and, in Augustine's account of her creation in Book 2 of his *De Genesi contra Manicheos*, Adam says "she will be my adjutant" (*haec erit mihi adiutorium*). In all the accounts, *mulier* is named Eve only after the Fall (Gen. 3: 20), when Adam's judgment has proved to be as poorly directed as hers. Augustine's account of the Fall, unlike others more concerned with punishing rule-breaking, turns exactly on matters of desire, will, and choice involved in the humans' rational judgments. On the woman's naming in the two Genesis accounts, see Conybeare 2018.

must "freeze the moment," "stop the action" for however brief an interval. This fact, itself an observation about an act of *recollecting* as well as an action of measuring, is a basic psychological observation even now (consider how we perceive what seems to us to be continuous motion in films).[22] These two elementary principles of human *mens* (*meminendi* and *metendi*, remembering and measuring) understand things not only in space but perhaps even more principally in time.[23] Because the natural cosmos is essentially change and movement, it is as much temporal as it is spatial. Movements unfold in space-time and can be measured only retrospectively, when they have been stopped (however arbitrarily) and are delimited by the two points – a start and a finish – that Euclid defines as "a line." We still measure speed and distance in units of time: light-years, miles per hour, and so on. The distances of medieval journeys were often expressed in terms of days travelled.

Throughout this book, readers need to be mindful that the mathematical and also social vocabulary that my sources use must be understood with their historical meanings, which are often less precise than the meanings they may have now. The logic governing their use is itself fundamentally different from ours, for it does not so sharply separate and compartmentalize reasoning as "linguistic," "ethico-social," or "mathematical." Instead there was assumed to be a basic connectedness among all the modes of human understanding, among physics, medicine, and rhetoric, politics, ethics, and the movements studied in *mathematica*, including all rhythms whether of the stars, human lives, or of speaking and singing. This has been called an "organic" or comprehensive logic rather than the mechanical, categorized, and "algorithmic" logic of the western Enlightenment.[24] The organic model is succinctly expressed by the

22 See Berthoz 1997 and Flash and Berthoz 2021. Though the latter announces that it is about mental geometry and "artists" it has no discussions of any of the arts of the classical trivium, which is a pity, since *geometria* is so deeply embedded in language arts, especially Dialectic and the rhetorical accounts of Invention. Serres 1982 discusses "The Origin of Geometry" as rooted in speaking, especially in dialogue, both actual and mental (a thoroughly Augustinian idea). Sounds are vibrating pulses of air, which we hear when their movement reaches us: in the first Genesis account, God creates in both light and sound. I discuss these matters in detail in Chapter 3, and see also Carruthers 2022.

23 The "Big Bang" and associated physical theories of relativity and quantum motion also privilege time, in sometimes remarkably similar ways. Rovelli writes that "the best grammar for thinking about the world is that of change, not of permanence. Not of being, but of becoming … the world is not a collection of things, it is a collection of events … a network of events" (ch. 6); see also Penrose 2011.

24 I thank Dr. Howard M. Robinson for some clarifying conversations about this matter, the Enlightenment mind–body "problem" (precursors of which begin to

twelfth-century master, Thierry of Chartres, commenting on Boethius's "On the fundamentals of arithmetic" (though he is by no means the first to express the idea, often attributed through the centuries to Pythagoras):

> quia sicut unitas est complicatio numeri, et numerus explicatio unitatis, ita quoque punctum est complicatio omnis magnitudinis, et magnitudo est explicatio puncti.
>
> [For just as unity is the enfolding of number, and number the unfolding of unity, so too the point is the enfolding of every magnitude, and magnitude is the unfolding of the point.][25]

A line can be drawn from its initiating point either *recte*, "straight, directly," or it can bend in a circle as it is drawn around a starting-off point. Straight lines move moment by moment until they may join with another straight line moving from a different direction. Their joinings form angles (which are measured in degrees, since they ultimately are abstracted from a spherical shape, the global Earth). This initiating angle (right, acute, obtuse) determines the other angles of a delimited (and thus measurable) triangle. All polygonal forms, including cubes, can be measured from their angles as sets of triangles that can then also be comprehended in terms of their linear bounds or *termini*: thus, logically (and theologically) all terrestrial shapes can be analyzed as sets of prismic shapes, and in this way their forms can be said to reflect multiple "trinities," but in a manner that accords with (darkened) human reasoning.[26]

---

emerge at least in fourteenth-century scholastic rationalism, with Burley, Ockham, and their peers, and indeed even before, in Roger Bacon).

25 Thierry of Chartres, "Commentary on Boethius' *De arithmetica*." Quoted from Albertson 2014, p. 129. Albertson's chapters (pp. 93–139 esp.) on the development, via Bernard of Chartres, of Thierry's Boethius-and-Augustine-implicated ideas concerning divine mediation and creative Folding need to be read carefully; they bring together "the Boethian-Peripatetic model (*theologia*, *mathematica*, *physica*) and the Augustinian-Stoic model (*ethica*, *logica*, *physica*). Rather than simply selecting one, Bernard cannily coordinates them, such that the Boethian *mathematica* (the quadrivium) defines the content of the Stoic *physica*. Physical science thus amounts to mathematical analysis of the sensible world" (pp. 101–2). Modes of human understanding and cosmic Being model the same created cosmos but in ways that understand it as *complexio*, Many "folded together" and expressed in many ways (*modi*). Latin *modus* basically means "measure" and also "means of understanding," in this way uniting Arithmos and Logos.

26 See S. Murray 2014, esp. pp. 37–46. As he observes (p. 45), "The triangle was thus the master figure in the reconciliation of the megadata of the real world with the mental architecture and theoretical comprehension of the beholder." See also Bork 2011, pp. 1–27, who stresses the fractal nature of the geometry of Gothic architecture, its fundamental fractals being quadratures and polyhedrons built in various

The Earth's shape was commonly represented in theological diagrams of the cosmos as *cubus*, a quadrilateral formed from two triangles – though nobody in western cultures ever believed that the globe (Lat. *globus*, by definition a round ball) was actually a cube: as John Donne wrote, the "round earth" has only "imagin'd corners."[27]

A circle (and a sphere) are described as mathematically *perfectus* when the line that encircles them is completed, in that the line that traces a circular movement has revolved about the central point (of the compass) and now has returned to its start, completing a full circumference (the action of *circumferens*, "carrying around"). The length of this line can now be measured using the tools of human reasoning. But neither the circumference nor the area it encloses can, in divine terms, be *perfectly* measured by human reasoning. This is the Archimedean problem (and theological mystery) of "squaring the circle" which results in the uncalculable number "pi" (π).[28] It encapsulates the theological belief that human reason is constrained and incomplete in its very nature. As Calcidius wrote, God is perfectly complete; then "in the beginning" Divinity (which in itself is dimensionless, timeless Point) unfolded,[29] dilating outward from Itself to form the cosmos, its own Being expressed dynamically through animating sound and light curving spirally all the while to mirror itself in the round globe that is the cosmic center, from which single fixed-perspective humans must observe and try to measure the temporal-spatial motions surrounding them. Though Earth may seem to us to be the still center point of an otherwise spherically moving cosmos, that is only because

ways from triangles; and Shelby 1972, an overview of how medieval masons learned their geometry (not much directly from books).

27 See Obrist 1997 and 2004. The convention of representing the earth as a cube within a sphere derives from ancient cosmic wind-diagrams, as Obrist explains. I discuss this again in Chapter 3. The quotation is of Donne's Holy Sonnet 7, line 1.

28 For those readers who may have forgotten their elementary Euclid, the formulae for calculating (measuring) circumference and area both involve π ($2\pi r$ and πr^2). Note that π is now called an *irrational* number, "irrational" because inexpressible in the terms of the created cosmos's "rational" harmonic schemes. Another such number is the square root of 2, which is the measure of the diagonal drawn transversely, corner to corner, across a square of equal sides. These two inexpressible measurements were much cited as demonstrating the ultimate incommensurability between divine Plan and human reasoning ability. Grattan-Guiness 1996 points out that Euclid himself called the circle-squaring problem "*alogos*" ("unsayable" and "plan-less"), its inexpressibility in human terms marking the limit between Eternity and temporal, moving Creation.

29 Notice this idiom and my use of a temporal conjunction – the logic of a point-like *principio* is embedded in English and other languages still. Sorabji 1983 is a most readable account of pre-modern ideas about time, continuum, and their relationships.

of our Earthly perspective. The Earth's nature is also one of constant change, movement. And the nature of movement is continuous, as one event "unfolds" into the next, even when those motions bend in the curves, spirals, and spheres of cosmic creation. These expressed spheres, moving constantly, are thereby also not perfect, for their ending is not identical with their beginning. If it were truly perfect, the cosmos would never change or move, for it would have no impelling, moving desire to be completed and fulfilled, an attraction (Latin *ad+traho*, "draw toward," "attract") that is the motivizing energy of this universe, both physical and ethical.[30]

The imperfect cosmos moves "spiritually," *Timaeus* says. In standard hexaemeral commentary on Genesis 1: 2, this spirit (*pneuma*) is identified with the Holy Spirit, "spiritus Dei" (or *anima mundi* or – later – *Primum Mobile*), the dynamic, energizing breath of God, which "moved (*ferebatur*) over the waters" shaping the temporal cosmos through God's initiating spoken command, "Fiat lux." The movement (animation) traced by this creating breath is described geometrically, in Platonist thought, as spiral, the observed movement of life-giving human breath also (best visible on cold days).[31] Latin *spiral* in this context is the word for a breathing-hole or blow-hole (such as volcanic steam vents from the Earth), from the verb *spirare*, "to breathe." A Latin homonym, *spiral* (directly borrowed from Greek σπεῖρα), denotes a "coil," the way a ship's ropes are kept on deck ready for tossing out in a straight line. In his discussion of this basic figure, Eric Palazzo demonstrates how a common medieval iconography shows the movements of natural Creation in spiraling forms, perhaps from having homonymically conflated the two Latin words (which isn't possible to do with Greek *pneuma* and *speira*), perhaps influenced also by Calcidius's account. (A different Latin noun, *spira*, denoted a spiky, reed-like plant, and in medieval Latin is the word for a church "spire": there was much invitation in these words for image- and word-play.)

30 Readers should keep in mind that the heavenly bodies, both the "errant" and "fixed" stars, are affixed like gem stones in a ring. Thus the outer planets (Mars, Jupiter, Saturn) seem to "wander" in a retrograde motion only relative to our perspective from Earth looking out toward them. We see them *only as we are able to see them*. The sphere of the stars is at such a vast distance (even though its motion is the swiftest) that humans cannot perceive or measure it (though some efforts were made to calculate the length of the Cosmic Year, as it is sometimes called). A welcome account of pre-Ptolemaic and early medieval explanations for retrograde planetary motions (including some heliocentric ones) is McLeish and Garrison 2021. This unbridgeable gulf between human knowing and divinity, thus physics and theology, lies at the heart of Galileo's controversy with the Inquisition: see Drake 1980.

31 Palazzo 2020. I discuss this idea further in Chapter 4.

Human reasoning also always moves among its materials, in pathways that can be described and pursued geometrically, like a journey among places. This is exactly what is meant in human composition by the "topics" for any argument. Our subject matters are not a jumble of content but are always inventoried and positioned by us in relation to one another, and their positions in any particular argument will change; "inventory" is another word deriving from *invenire*. This quality enables a variety of locations and relationships among our subjects, degrees of likeness and difference, comparison and contrast (to invoke a particularly useful, and sometimes even delightful basic method). "Method" derives from Greek *hodos*, a "path"; a method of reasoning isn't a choice between irrevocable binaries, but more like choosing among pathways that can meet up at various angles, perhaps even at later stages of the journey, more like a labyrinth than a clearly terminated line.

It is our crafty *shaping* of our investigating methods and invented materials, the organization or *Dispositio* (as it is named in Rhetoric) that we will construct among them, that is crucial to their intellectual usefulness. This is an important way in which the craftsman God and human crafting, however humble by comparison, can be described as analogous. It is also why this book, though about the art of human inventing in rhetoric and meditation, will pay such close attention to the essential role of geometry in crafting with words. All composed plans – visual, verbal, and musical – through the articulation of their parts (in other words, their syntax) invite specific kinds of internally related movements, and thus imply certain particular logical relationships. The ways in which a diagram can take you from one of its "places" to another, how its "topics" are disposed, what sorts of movement are designed into it – in that lies its investigatory and inventive power, and – inevitably – also its limitations.[32] Scholars now tend to classify diagrams by their audiences (school diagrams, preachers' diagrams, diagrams for surgeons, diagrams for singers, etc.) and by their content (virtues and vices, divisions of knowledge, human anatomy, cosmology, orchestrations, etc.). But medieval classifications most commonly denominate shapes – wheels of this and that, trees of this or that, ladders, hands, towers, cherub wings, spheres, and more. The implied internal motions of these basic shapes can be linear, compartmental, genealogical, circular, spiral, diagonal; they can employ a linear geometry, a spherical geometry, or even employ bending and stretching movements so complexly interrelated that they are best considered together geometrically as a topology.[33]

32 I have discussed the pervasive pre-modern idea of composition as *ductus* in Carruthers 2010.

33 Saiber and Mbirika 2013 is the result of a collaboration of Dante scholars and mathematicians who attempted with the aid of a computer to visualize as a ring-knot

What constitutes a recollective search/investigation is discussed by Augustine, and his characterization of recollection in *De trinitate* Book XI remained key. There are at least two kinds of reconstructive recollection. In one kind, you can recall now that you once made a memory of something but you no longer can view the content. This is the "I know I know that but I just can't recall it" situation, your present remembering that in the past you made a memory of something you experienced.[34] Your task then is to discover again your memory trace of that content (but, notice, not the initial content as you first cognized it, for that affect – or "motion" in Aristotelian terms – no longer "moves" you). You do so by reasoning sequentially from a memory-trace that you now can locate until you find the forgotten notion that has been *oblitus*, "covered over" and undiscoverable, but that you now wish to recollect into the present. In order to use it for your present thinking, you must recreate it now with the aid of a newly fashioned image.[35] The usual example given in scholastic writings is to try to remember what you did four days ago by starting from experiences you can now recall (perhaps of what you did yesterday) and from there reconstructing two days ago, then three days ago, and finally four days ago. Or you can proceed in the opposite direction, from four days ago to yesterday, if that is what you need to do. You have great flexibility in how to proceed, so long as your association chains are reliable.

Another – and much more interesting – kind of rememorative investigation of one's notions and concepts occurs when one seeks to understand matters that one has never experienced oneself. A common Scholastic example, derived from Averroes, is how a person who has never seen an elephant can yet conceive of one from having it described in words only. One must investigate one's conceptual treasury also in order to think through matters that can never be directly cognized, such as abstract ideas like Triangularity. To

---

topology the final figure for the Trinity in Dante's *Paradiso*, described in the text as an interlocked wheeling figure of *tre giri*, "three wheels"; it appears to the Pilgrim-Dante to be both one and three, motionless (as Eternity) and at the same time appearing to have both dimension and movement, for otherwise the poem could not speak to us of the figure at all. The most satisfactory topology, they suggest, is known as Borromean rings. The authors admit they were not wholly successful in their search, but their effort pays tribute to Dante's remarkably precise crafting imagination.

34 Note that medieval psychology did not sharply distinguish, as modern psychology does, between so-called "semantic" and "episodic" types of remembering; see Carruthers 1998.

35 Accuracy is not a matter that was understood in anything approaching its modern meaning of an "exact replica" of an original event or work. I discuss this in more detail in Chapter 2.

consider "What is triangleness?" Aristotle says, one must imagine a triangular shape, understanding that one is not defining this particular triangle-image but, rather, using it to understand the idea of Triangularity.[36] Augustine used other examples, such as: I never saw St. Paul but I can fashion a working mental image of him by recalling and reimagining my notions of "a man" and (perhaps) "baldness" to help me comprehend my thought of "St. Paul." (Frequent later advice is to use a graphic image of "St. Paul" that you have seen in a church or book.) Similarly, even though I have never been to Ephesus, I can imagine what the city may be like because I have a memory of my experience of another city (for Augustine, it was Carthage), and so I can comprehend "city of Ephesus" as a concept. And I can also conceive of things that I could never experience and that do not even exist, because I can combine my recollected concepts in new ways. Augustine's example is a black swan (in ancient logic this was a standard example of a categorical contradiction). Since this is an ontological contradiction, in nature it cannot actually exist. But because I can recall my notions of "swan" and also of "black" and visualize them together, I can indeed imagine and understand the idea "black swan" in the abstract (as we now say) as a generalized metonym for logical impossibilities.

We now call this using our imaginations, and revere the Imagination as our greatest creative ability, by some enthusiastic Romantics claimed even to be independent of both memory and reason. But – as I have often observed in print – in medieval psychology, including both Augustinian and Scholastic psychology, what we call "creative imagination" was analyzed as rememoration, a conscious investigation of our conceptualized experiences, which are reconstituted for intellectual use by our recollecting and reimaging abilities. For this reason, *investigatio rememorativa* was regarded as the all-important bridge between our sensory experiences and our theorizing intellect, between body and mind.

36 See Aristotle, *De memoria et reminiscentia* I (450a): "It is impossible even to think [Gk. *noein*] without a mental picture," citing his previous discussion in *De anima*, III.vii (421a): "Now for the thinking soul [Gk. *psuche*] images [Gk. *phantasmata*] take the place of direct perceptions.... Hence the soul never thinks without a mental images [Gk. *phantasmatos*]"; ed. and trans. W. S. Hett for LCL. Hett translates Aristotle's term *phantasias* as "imagination," and it is clear in Aristotle's account that "imagination" is not a separate organ within the brain, the way the Enlightenment and Romantic poets and philosophers came to do, but a mode of intelligent activity known as *psuche*, the soul thinking. Aristotle's analysis was important in medieval philosophy (including that of Augustine) to establish that the human soul is a single animating energy for the body that is active in a variety of ways (rather than each action requiring a different "soul").

In procedures of thinking, the fundamental reasoning pathway that each diagram's shape offers affords us "a framework of recognition."[37] This gives priority to *how you look* before *what you see* – it is similar to what is now called perceptual learning via processes of pattern-detection and pattern-making. It is not a passive, wholly receptive matter at all. Many people now are familiar with it from the Duck-or-Rabbit inkblot psychology test. A group of DNA-genome researchers whose work was resulting in very large data stores decided to bring in an artist to teach them how to expand the meager set of patterns with which they were attempting to shape and analyze their random-seeming data, so that they could visualize them using a more varied stock of possibilities.[38] Diagrams help train our perceptions, our abilities for seeing/making patterns and therefore finding relationships in our world. The users' judgment is involved, their "free choice," *liberum arbitrium* constituted of both volition and reason. For *ratio* means "plan" and "pattern"; *voluntas*, "desire" and "choice," and different patterns embody and invite differing perceptions of the same content and require us to judge not only what but *how* we are "seeing things" in our materials.

To see the same matters in different formations, I submit, does indeed change – and, I think, frees up – how we are able to understand them. Design is not just an ornamental vestment that can be cast aside for the real stuff, the chaff for the wheat, as it were. (I think we all know this, but we sometimes act as if we did not, when in our own classifications we lazily separate the content out from its organizing pattern.) A common feature of the medieval lexicon used for geometry will likely be particularly odd and counter-intuitive to us (unless, as scholars of manuscripts, we have learned to visualize their drawn shapes within medieval conventions). The simple planar shapes – square, circle, triangle – when one encounters them in diagrams, especially as structures for topical Invention, should first be understood and envisioned for use in thinking as though they represented solids: cube (cubit), sphere, wheel, prism; or to use terms even more common in medieval geometric thought, "quadrilaterals," "roundnesses," and "triangularities."[39]

The following chapters explore these issues in much more detail.

37 J. Franklin 2000.

38 A popular account of one such effort is Carey 2015.

39 See DMLBS s.vv. *quadrus*, *rotunditas*, *triangularitas*. There are a number of related forms commonly used to denote such shapes; see an eleventh-century drawing of several such shapes in Figure 2. Grammatically, it is best to think of these terms as adjectives describing the names of the many particular things from which the mathematical abstractions are derived and defined.

Chapter 1 begins with an analysis of Nicolas of Cusa's dialogue "On a bowling game" (*De ludo globi*) and considers the important role of game-playing and pedagogical disputation – "arguing from both sides" – in the practices of invention, meditation, and philosophical exploration. It then considers the conditions thought best conducive to creative meditation and envisioned composing: dislocating from one's ordinary routines to a place of solitude; approaching the task in disquiet of mind and anxiety of spirit; being perplexed and unsure of one's understanding of the matters one is hoping to consider (frequently one's surrounding light is dim and flickering, like candle light); and finally being able to see with the mind's eye (*oculus mentis*) the ordering shapes, often in detailed, pictured form, of the composition one is planning. I then consider some examples of such meditational beginnings, including Bonaventure's withdrawal in solitude at the start of his *Journey of the Mind Toward God* and Boethius's description of his beginning vision of Lady Philosophy in his *Consolation of Philosophy*, including paintings of this scene of Invention in various manuscripts of this work, and also of Hildegard of Bingen's envisioned *Scivias*, her "ways to knowing" God. The chapter ends with an analysis of one of the best descriptions we have of an act of creative meditation through dialogic argument and self-persuasion, Augustine's so-called "conversion" (though "persuasion" is a better term) to a celibate life as recounted in his *Confessions*.

Chapter 2 concentrates on the relations between reasoning and envisioning, focusing on the rational invention method of *Topica*, employing topical "locations" and their relationships, and using geometric figures as a primary tool for forming a composition and persuading others to confidently trust and believe what you say. (The subject of Topical reasoning immediately preceded that of rhetorical Invention in the liberal arts curriculum, before Dialectic and Rhetoric came to switch places during the later thirteenth century.) Beginning with Aristotle's observation that philosophizing begins in wondering, it focuses on the relationship of imagining and reasoning not in Scholastic philosophy but, rather, in Richard of St. Victor's guides to contemplative practice, the so-called *Benjamin minor* and *major*. The Victorine texts differ from the later university ones in emphasizing the limitations of human reasoning and the importance of bodily passions and ardent desire, and of experience-derived imagining in continual searching for religious truth. In this light, I also consider the teaching of John of Salisbury about synergies among reasoning, passionate bodily feeling, and imagination in his *Metalogicon*, along with other related twelfth-century Cistercian-influenced texts, including *De spiritu et anima* and some work by Clarembaldus of Arras. The concept of "high fantasy," Dante's *alta fantasia*, is but one aspect of this Victorine-Cistercian emphasis. The chapter concludes by considering the roles for inventive thinking of joy, dancing, and of constant shaping and reshaping in Richard of St. Victor's account of human thinking in *De contemplatione* (*Benjamin major*).

Chapter 3 focuses on late ancient and medieval accounts of the complex roles of geometric figurations (*geometricalia*) in all forms of starting the process of creating, both human and divine. It augments the concepts briefly discussed in this Introduction, paying particular attention to the basic notion of space-time unfolding and unscrolling, moving from its initiating point outward into the various actualized forms of the changing, created cosmos. And it pays detailed attention to medieval concepts regarding the shaping of natural forms, and to the elementary definition (found in Cassiodorus and Isidore) of *mathematica* as the science of abstractions fashioned by reasoning human beings from the forms of the creatures around them in order that they could measure and thus understand the natural cosmos as fully and truly as their natural mental constraints enabled them (the perceived etymological link in Latin of *mens* with *mensura* derives from this idea). God brought forth the physical cosmos (Something from Nothing, as the Hebrew says) in Time simultaneously with Number, Measure, and Mass; in order to understand this creation, human beings invented the disciplines of *mathesis*, that is, number and geometry. Time is essential to the processes of natural creation, since all change consists of movements perceived within passing time: God's creating energies are measured in Days, as Genesis says. As an example of using a geometric shape to structure a complex meditation, the chapter next moves to consider Alan of Lille's unfinished composition "On the Intelligible Sphere," whose initiating text observes that "God is a comprehensible sphere whose center is everywhere, circumference nowhere." Through this text and other similar self-consciously rhetorical meditations (Alan's text is now generically classified as a *sermo*) the connection between geometry and oratory apparent to Quintilian and so many others becomes far more evident and playfully creative, a mirroring of divine creation by human art. The chapter then focuses on relationships among sound (music), mathesis (number and geometry), and grammar as defined chiefly by Boethius (on mathesis and music) and Priscian (whose work on Latin grammar asserts strongly that language is essentially made from sounding voices, *voces*, of which the expressed lines (*grammae*) of written scripts (*litterae*) are the figures whereby humans may reconstruct those voices from a silent page. Because every sound sets in motion a particular corporeal vibration, a melody can effect in humans a particular passion/affection/mood (three concepts synonymously used in medieval languages) just as spoken poetry does; thus music can directly shape human feelings and moods, a characteristic particularly emphasized in ancient medicine, as well as within the arts curriculum and in the various aspects of Moral Philosophy, for when received within the body sound arouses desire and thus has ethical character. The chapter concludes with a consideration of how the notion of Time as continuing movement, infinitely unfolding in the present from past towards future, models basic notions of human and revealed history, via the Pauline concept of "the fullness of time."

Chapter 4 examines some examples of how these various geometricals were used to shape particular texts, and the varying logical relationships made possible when the same content is located within different figurations. It begins with simple cubes (*cybus*),[40] quadrates defined by linear angles and made up of two triangles. My example is the basic Square of Opposition as presented dramatically by Martianus Capella in the figure of Dialectic. Though she is one of the language arts, her discourse and her demonstrations are entirely geometric in form. She embodies the Calcidian (and ancient) doctrine that logic (as Logos) and geometry are both measuring instruments and together are the foundation of confident belief and understanding and of the creating forces of harmony (music) in nature and society. Quadrate geometricals commonly used for biblical meditation and exegesis are Noah's Ark and the Exodus Tabernacle and Covenant Ark. As shown in manuscripts, these figurations are derived from and need to be imagined and understood as representing solid forms. Several conventions for depicting them in manuscripts, notably with a double-lined frame, indicate this. As basic tools for seeing, moving about in, and using them to construct one's thoughts, such quadrate figures have a long, distinguished career in medieval education at all levels, from novices to university scholars and adept mystics. I discuss in particular the shape known as an "embedded square," used notably in some exegetical paintings of the Heavenly City to indicate its three-dimensional form.

The chapter then turns to some other major families of useful shapes; first to circles, semi-circles, and other round structures such as wheels, and briefly considers other examples used primarily for philosophical contemplation and

40 A medieval *cybus* or *cubus* (translated as "cubit" in the English Bibles) is not just the solid form defined mathematically now as a cube. Not only can it include any *forma quadrata*, such as what we are taught to distinguish as various rectangles, but what we would identify as a planar "square" is often labeled *cybus* or *cubus* when drawn in *geometria* manuscripts. Much nineteenth-century exegetical scholarship went into defining the measure called a "cubit," and biblical scholars seem agreed that there were different precise quantities, but measured around 18–20 "inches." The Bible itself distinguishes between the "Egyptian" and the "Hebrew" cubits, the "Egyptian" being longer. As is the case with ancient and medieval units, their uses and quantities were at best only locally standardized, so the quantity of an "inch," defined as the length of the upper joint of the thumb, is also variable. A helpful discussion of these matters is Stone 2014; see also Lugli 1996, a masterful historical overview of the basic human need for measurement.

meditation such as winding towers and round world maps (*mappae mundi*). Then it considers figures such as the ubiquitous branching and elevating forms of various tree diagrams, ladders, and steps. These are among the earliest and most widely used of all. Connecting all these various forms is their logical and compositional utility, more so than any mimetic qualities they may also have. Their shapes invite use, and because they are so various and so ubiquitous, they also underscore how the same subjects may profitably be arranged in different forms to make different connections and see particular subject-matters differently. I then discuss one example of this, considering how simple linear journeying, circling, and climbing structures invite very different understandings of the relationships among the Liberal Arts, and their joint family relationships to Philosophy and Theology.

Finally, the book examines briefly the use of several geometrical forms by Richard of St. Victor in the prose-poem that summarizes his "Mystical Ark" (*De contemplatione*). Called "Some Allegories of the Tabernacle of the Covenant," in it he counsels his Victorine brethren on how they may use the several allegorical possibilities he has linked to the different shapes of the Tabernacle's furnishings for their own meditating, composing, and preaching purposes.

1

Syntax and Geometry

Curiosities and Games

One of the odder texts of the late Middle Ages is a meditational and pedagogical dialogue, an education for princes [*regimen principis*] concerning Godhead, Creation, the nature of free will (divine and human), and the virtuous life, structured in terms of playing a "game of bowls" [*ludus globi*]. Its author, Nicolas of Cusa (Nikolaus Cusanus, 1401–64), was a learned cardinal-bishop, but not a university-centered academic. His career was spent in papal administration, moving easily between Germany and Italy during a notably contentious period in the western Church. He was also possessed of one of the most inquisitive philosophical intellects of the notably inquisitive fifteenth century, thoroughly learned in medieval philosophy as well as in the Ancients. In this Dialogue, composed (1462–63) towards the end of his life, he imagines conversing with two young princes of Bavaria, John (in Book I) and Albert (in Book II) on a field laid out for playing bowls. Cusanus imagines himself seated alone on the sidelines, tired out from the game he has taught to them, the one which the younger princes are now actively playing. Prince John asks him to explain the rationale of this new (to him at least) delightful game [*ludus iocundus*]:

> *John*: All of us are fascinated with this delightful new game – perhaps because in it there is the figuration [=*figuratio*] of a kind of cosmic contemplation that we beg to have explained.
>
> *Cardinal*: Not a bad move for us to make [*non male movemini*], various sciences have their instruments and games: arithmetic has its rithmomachy;[1]

[1] Exactly how the *rhythmimachia* referenced in the text was played is uncertain, but it was a numerical game using pieces of different values and shapes moved about a checkered board, rather like chess. The earliest known reference to it is ninth century, in a work from the monastery of Reichenau, and it has been suggested that it was invented there (whatever that may mean) as a means of teaching arithmetic to novices. A twelfth-century account of it in Cambridge, Trinity College MS

music has its monochord; and even the game of chess does not lack for concealed ethical knowledge [*mysterio moralium*]. I believe that no well-devised game [*honestus ludus*] is inwardly void of teaching and learning [*disciplina vacuum*]. For example, this delightful practice of bowling, it seems to me, can represent to us no small amount of philosophy.

John: So say something about it, we ask.[2]

What follows is both an ethical *speculum* for princes (both of them) and an *alta speculatio* on divinity and cosmos, a macro-microcosmic meditation on the virtues required for governance of both cosmos and state.[3]

The game played in his *Dialogue*, Nicolas says, is his invention: "I thought about inventing a game of wisdom. I considered how that game would have to be structured."[4] The playing field he describes consists in nine circles

---

R.15.16 has a drawing of the board, and also mentions its likeness to chess; see Burnett 2010, essay III (first pub. 1996), 244–52, at 247 esp. There is a long-standing, influential Neoplatonist tradition within Christianity of meditation on divine matters (*alta speculatio*) using number sequences; see Albertson 2014.

2 Nicolas of Cusa (Nikolaus Cusanus), *Dialogus de ludo globi*, Bk. I, 1–2. All Latin quotations are from the Paris edition of 1541 as edited by H. G. Senger (1998). While I consulted the translations of J. Hopkins (2001) and P. M. Watts (1986), the translations are mine.

> "*Ioannes*: Admiramur omnes hunc novum iucundumque ludum, forte quia in ipso est alicuius altae speculationis figuratio, quam rogamus explanari.
>
> *Cardinalis*: Non male movemini, habent enim aliquae scientiae instrumenta et ludos, arithmetrica rhythmimachiam, musica monochordum. Nec ludus scacorum caret mysterio moralium. Nullum enim puto honestum ludum penitus disciplina vacuum. Hoc enim tam iucundum globi exercitium nobis non parvam puto repraesentare philosophiam.
>
> *Ioannes*: Aliquid igitur rogamus dicito."

3 It is tempting to understand, in Cusanus's mention of educational meditations using a chess game, a specific allusion to the Dominican friar Jacopo da Cessole's "Game of Chess" (ca. 1300), a widely disseminated ethical meditation in the *regimen principis* genre, using a chess game as its structuring device. But there is no proof of this. Chess is also the basis of a late medieval (ca. 1377) French *regimen principis*, called *Les eschés amoureux*, or *Les Eschéz d'Amours* (its two manuscripts have slightly variant titles). Chess seems particularly well suited to structure a dialogue for those trained in *disputatio*. See D. del Puppo 2012.

4 *De ludo globi*, Bk. I, 31: "Cogitavi inuenire ludum sapientiae. Consideravi quomodo illum fieri oporteret." On the imaginary game, see Heller 2012. Heller makes the point that the dialogue genre is also analogous to a game of catch, as one speaker catches the ball from the other and throws it back, movements Cusanus exploits as much as he does the imagined movements of the uneven ball bowled on the playing field. As Heller writes, "Cusa's metaphors are not statements. They

concentric about a tenth, at the center. Based on a bowling game something like modern lawn bowls (rather than American-style pin-bowling), each player throws a solid wooden ball (*globus*) toward the central circle. However, the ball is crafted in such a way that a scoop has been taken out of one hemisphere, so that the ball cannot roll in a straight line, but only elliptically towards (or sometimes even away from) the target center. The ball's impetus and consequent motion is specific to each throw, and thus its movements will vary according to the physical and mental coordination, skill, and experience of an individual player. Most importantly, as "the Cardinal" (Cusanus) stresses, it is a game that someone can get progressively better at, the more practice and learning they acquire – one is not just a prisoner of the physics of the ball and the lay of the field. Since the game is imaginary, one's mental eye needs to be especially well trained and acute; as the Cardinal says to Prince Albert at the start of their dialogue: "It is necessary that you open your mental eye and elevate its visual ability fully, so that you may more powerfully see rather than just hear the things I am about to say."[5] For, as imagined, this game of bowls is a highly interactive occasion of learning, as we now likely and truthfully would say.

The first book of *Dialogus de ludo globi* focuses especially on the concept of *rotunditas*, "roundness." In the contemplative exercises of medieval *alta speculatio*, the "cosmic contemplation" to which Prince John alludes, figures of circles and spheres predominate, for evident reasons: a sphere is the geometric shape that seems best to structure reasoning about God's creating expression into and as all the forms of the universe, a divine cosmic force pressing outward through all creatures. The exercise has a long tradition. Alan of Lille's so-called "sermon on the intelligible sphere" (from the 1160s) is also a meditation on Creation, beginning with the topic sentence, "Deus est spaera intelligibilis cuius centrum ubique circumferentia nusquam [God is an intelligible sphere whose center is everywhere, circumference nowhere]."[6]

are complex puzzles that cannot be solved, or that could be solved in many ways, depending on what one calls 'a solution'" (473). The thought-provoking and thought-structuring role of play and game in ancient Greco-Alexandrine mathematics is the subject of two helpful studies: Netz 2009, on numbers and poetry, and Netz 1999, on proof and diagrams. The necessary role of play in creation is also a major theme in the biblical books of Wisdom and Proverbs, both greatly influenced by Judeo-Alexandrine thought; see Carruthers 2013, esp. pp. 16–44.

5 *De ludo globi*, Bk. II.61: "Oportet autem, ut mentis oculum aperias et visum illum totaliter eleves, ut quae dicturus sum potius videas quam audias."

6 Edited by W. P. Wetherbee, *Alan of Lille: Literary Works* (2013), pp. 1–17, at 2–3. Though this saying is attributed by many later writers to the unknown "Hermes Trismegistus," Alanus gives it no attribution, and, following Dronke 1974, p. 145, Wetherbee agrees that it may well have been his own wording. Cusanus also uses

Composed some three hundred years before Cusanus's bowling game, this work uses geometric forms in a similar way, to organize and shape a rhetorically brief, theologically compressed composition about the cosmos, the Trinity, and the soul.[7] The content of Alan's *sermo* is Neoplatonist, Boethian, and also is influenced immediately by the twelfth-century metaphysics of the Chartrians and Victorines, as indeed is Cusanus's *Dialogue*.[8]

My present book is about weaving narratives from all sorts of stored materials (including the common narratives bonding societies, the *dicta et facta* of a culture, as the Romans spoke of them). So was my earlier study of monastic meditation practices, *The Craft of Thought*. In this book, however, I want to focus more particularly on how certain methods of Roman rhetoric – known through Cicero, Quintilian, the various later Latin rhetoricians, Augustine, and Boethius[9] – provided much of the seedcorn for monastic meditation practices

---

the adage in his works. Meditation on cosmic relationships in terms of a sphere (or a variant thereof, such as Hildegard of Bingen's cosmic egg) is very much older, associated in the West with varieties of Platonism, the earliest Latin example being that of Ambrose of Milan, composed in the later fourth century.

7 Characterized by Wetherbee as "strange … imaginative and occasionally bizarre" (p. xii), this *sermo* is thought to be contemporary with Alan's better-known *De planctu Naturae*, composed no later than the 1160s (see Wetherbee, pp. x–xvi). It is found in a single manuscript, and remained unedited until 1965 (by M.-T. d'Alverny, *Textes inédits*, pp. 297–306). I discuss it in much greater detail in Chapter 3.

8 Albertson 2014 discusses at length the important influence of Thierry of Chartres (and others of that Neoplatonist twelfth-century group) on Cusanus. On Thierry, Isaac of Stella, and other twelfth-century Victorine-Cistercian teachings concerning body–soul relationships, and the particular re-emphasis at this time on a macrocosmic–microcosmic model of the universe (which they inherited from late antiquity), see the excellent Introduction by McGinn to his collection of translated texts, *Three Treatises on Man* (1977), pp. 1–93.

9 Before the fourteenth century, the important rhetoric masters between Quintilian and Boethius included Julius Victor, Fortunatianus, Martianus Capella, and Boethius's near-contemporary Cassiodorus: translations from their Latin manuals are in Copeland and Sluiter 2009. But the most influential late classical rhetoric master for medieval Western Europe, both for his theory and as a practical model, is Augustine: see Murphy 1974, Kennedy 1994, C. Lanham 2001 and 2002, Cox and Ward 2006, Copeland 2014. In addition to the few written manuals that we still have, the bulk of meditation practices in the Greco-Latin Christian West, which were largely Jewish and Alexandrian in their origins and nurture, were taught through apprenticeship (for novices) and practiced only orally. The evidence we have for these practices comes indirectly in the surviving meditations and prayer texts themselves, notably the *Praktikos* and *Chapters on Prayer* of Evagrius, the *Lives* and *Sayings of the Desert Fathers* (translated to Latin from Greek) and the works of John Cassian. See (among other studies) Louth 1981; Leclercq 1982; McGinn 1991; Leyser 1994.

in the twelfth through the fourteenth century (and well after that) for contemplatives and poets both in and out of clerical orders, composing in Latin and in various European vernaculars. It is an unfortunate result of the manual-based, written pedagogy of academic rhetoric that Invention is so often described as though it were a distinct product that is then shaped up by the other four distinct rhetorical "parts" (Organization, Style, Memory, and Delivery), a separation perpetuated by its modern university course-title of Composition and Rhetoric. For, in actual practice, Invention is concordant and synchronous with rhetorical Dis-position (*Dispositio*), which is literally "putting into (coherently designed and clearly designated) places" the materials for any composition as one discovers and shapes them up; employing the same notion, Com-position also places things together in a synergetic complex. The governing notions in both terms are "movement" and "placement": Ordering is conceived of primarily in terms of the relationships among materials collected (*collectio*) in an in-forming space, and the varieties of movements (their in-forming logic and argument) that these gatherings can indicate and invite gives them their mutual coherence (syntax). In other words, complex Invention is always also a matter of Geometry.

Finding the Things We Say

When discussing the Liberal Arts in *Didascalicon,* his magisterial treatise on the monastic curriculum, Hugh of St. Victor writes of Geometry, that it is "the teaching of fixed measurements, and the depiction of shapes for contemplating, through which the boundaries of each and every matter can be made clear; or in other words, geometry is 'the well-spring of our sensory cognitions and the originator of the things we say'."[10] When I first came upon this statement I was greatly puzzled: what did geometry have to do with speaking

10 Hugh of St. Victor, *Didascalicon* II. 15: "[G]eometria est disciplina magnitudinis immobilis formarumque descriptio contemplatiua, per quam uniuscuiusque termini declarari solent. aliter, geometria est fons sensuum et origo dictionum." This brief definition is immediately preceded by one of Music as "diuisio sonorum et uocum uarietas. aliter, musica siue harmonia est plurium dissimilium in unum redactorum concordia. [Music is the division of sounds and diversity of voices/words. But also, music or harmony is the concordance of the many dissimilarities rendered into one]"; Latin text ed. Buttimer 1939; this and following translations of it are mine, consulting Taylor 1961. One should always keep in mind that poetic meters were taught as Music in the medieval scheme of Liberal Arts, poetry also being "a division of sounds and variety of voices," a definition that emphasizes the *sound* of poetry, not only its grammar.

and contemplation? I understood of course that Geometry was concerned with measuring forms that had immobile boundaries, in contrast to Astronomy, which dealt with measuring moving bodies: that distinction is commonplace in medieval definitions of the quadrivium. Nor, unfortunately, did I then take sufficient notice of Jerome Taylor's comment on this passage in his translation of *Didascalicon*.[11] For Hugh is quoting a maxim found in both Cassiodorus and Isidore of Seville – *fontes sensuum et origo dictionum* – but his quotation is not quite complete. Cassiodorus's complete dictum (copied fully by Isidore in his *Etymologiae*) is "topica … sunt argumentorum sedes, fontes sensuum, et origines dictionum [the Topics … are the seats of arguments, well-springs of our sense-based cognitions, and originators of the things we say]."[12] By Hugh's time, the maxim was commonplace, so well known that, without special attribution, Hugh could fold it into a much longer sentence.

My surprise lay in the context *where* Hugh cites it. Nor was I alone – it is so unexpected to us that it has been suspected as a scribal error or as simple ignorance of Euclid on Hugh's part.[13] Most Greco-Roman writers had applied the quotation in their discussions of the trivium, specifically to *Topica*, the method of placing arguments relationally within a coherently shaped structure, and which was taught in logic and rhetoric for analyzing and investigating materials by means of topical categories (Dialectic), and for finding and developing arguments for invention (Rhetoric). Topical reasoning always had a dual function – assessment (*ars iudicandi*) and investigation (*ars inveniendi*) – just as Dialectic and Rhetoric were considered two aspects of the same rational instrument, as the very same human hand offers both the "closed fist" of dialectic and the "open hand" of rhetoric, or, in Aristotle's model, are the strophe and antistrophe that structure a Greek chorus.[14] Grammar was

11 *Didascalicon*, Taylor, p. 203 n55; commenting on the phrase "fons sensuum et origo dictionum," Taylor notes that this Cassidorean phrase, though evidently displaced from its normal position in his *Institutiones*, is recorded in this location in all the many of Hugh's *Didascalicon* manuscripts; it cannot be dismissed as a scribal error.

12 Cassiodorus, *Inst.* II. 3.14.

13 M. Evans 1980, p. 34, characterized Hugh's statement as "extraordinary," mere "woolly enthusiasm" typical of a time before "proper Euclidian geometry" became a schools subject. An assessment of the limitations of Euclidian geometry in contrast to ancient (and medieval) ideas of time, location, and motion, written by a quantum physicist but for general readers, is Rovelli 2018. The rational limits of Euclid's theoretical *figurae* were also apparent to medieval writers – see Chapter 3, below.

14 As by Cicero *Topica* 6, and Boethius in *De topicis differentiis* III.2.2. The fist/hand analogy was best known in the earlier Middle Ages from Cassiodorus, *Inst.* II.3.2: "Dialectica et rethorica est quod in manu hominis pugnus astrictus et palma

the foundational assessment and analyzing aspect of this didactic scheme for learning the language arts, the tri-partite *Trivium* or *Logica*. As Cicero defined them in his *Topica*, a work often better known than Aristotle's in the Ciceronian, juridically oriented curriculum of practical rhetoric, the *loci* ("places," translating Aristotle's word *topoi*) are so called for they are like "seats" in which we can readily find, without error or distraction, those arguments we may wish to use for our own inventing/composing.[15] The word refers not to a formal argument so much as to the methods of reasoning that can render matters you want to be trusted and valued but about which the people in your audience may have doubts, to be worthy of their confident belief in what you are saying, and even to act on it, for good or for ill. These persuasive powers are also why audiences must judge the ethical character of the speaker, for a bad person may well give a well-crafted speech but not one to be acted on.

These logical "seats" each have their place in any completed structure. Towards the end of *Topica*, Cicero turns to geometry to demonstrate this. Summing up the various kinds of arguing, the *genera dicendi*, that he has defined and described, he announces that he has now completed his description, with examples, of all the types, and then names them briefly (arguing

---

distensa," quoting in turn from Varro's now-lost work on the liberal arts; it is discussed at length by Corbett 1969. Aristotle likens their relationship to choral antistrophe and strophe in the very first words of his *Rhetoric*. This same Cassiodorus text is cited by John of Salisbury, *Metalogicon* 3.9, during his discussion of *Topics*, which is its usual school-book position. Its Latin variations derive ultimately from Cicero's *Topica* as commented on by Boethius, a text widely known among medieval students. Boethius also wrote a separate logical work, *De topicis differentiis*, ed. E. Stump 1988. On Boethius's significant contributions to the subsequent curriculum for medieval dialectic, see Stump 1989 and Marenbon 2003. Reinhardt's copious notes to his 2003 edition of Cicero's *Topica* are indispensable. Aristotle's *Topica* was always known through its final placement in *Organon*, the Boethius-translated curriculum of logical analysis that came to dominate the university schools, especially in Oxford and Paris after the later twelfth century.

15 Cicero, *Topica* 7–8, ed. Reinhardt: "Ut igitur earum rerum quae absconditae sunt demonstrato et notato loco facilis inventio est, sic, cum pervestigare argumentum aliquod volumus, locos nosse debemus; sic enim appellatae ab Aristotele sunt eae quasi sedes, e quibus argumenta promuntur. Itaque licet definire locum esse argumenti sedem, argumentum autem rationem quae rei dubiae faciat fidem. [The finding of matters that are hidden is easy when their location is pointed out and labeled; just so if we wish to discover a particular argument, we should know the places, for these are so named by Aristotle as though they are the seats from which arguments are drawn. So we may define *locus* as the seat of an argument, and *argumentum* as the rationale by which belief may be constructed from something in doubt.]" All subsequent quotations are from Reinhardt's edition. The final translations are mine, after extensively consulting his.

from *definitio*, from *partitio*, from *notatio*, from *coniugatio*, from *contraria*, and so on). No seats of argument, he says, remain to be explored.[16] It was common in logical works to map the "parts" (*partes*) of a method of arguing onto a geometric form in order to show their internal syntactic relationships: one of longest-lived of these diagrams is the "Square of Opposition" (I discuss this in Chapter 4). Cicero seems to be imagining his structure as a square also. His list consists of sixteen types, occupying sixteen equal "seats" within the whole, like particulars within a family (he makes much of the derivation of *genus* from Latin *gens*, "family"). A square is easily divided into sixteen equal parts, a kind of "squaring" by fours. So instead of a Square limited only to necessary contraries he has imagined a Square for the whole set of generic forms (*formae*, his preferred word to translate Aristotle's *eidos*, "appearance, shape") for persuasive reasoning. One might call it a Square of Topical (and Typical) Argumentation, which includes the "contraries" set out on the Square of Opposites, but has room for a good deal more complexity and nuance by way of "customary" and "probable" argument, more suitable to the multi-generative "fountains" (*fontes*) of rhetoric than to the strictly binary "closed fist" of formal syllogism.[17]

Verbal composition (*ars inveniendi*) was practiced in the classroom through a series of exercises, the *progymnasmata*, that focused on various means and manners for making a finished composition, usually a forensic oration arguing (most often) the guilt of some criminal individual. One of these exercises was also called *topos* or "common places." The Greek teacher, Theon (writing a bit before Quintilian) defines the exercise of *topos* in a way that makes clear that it had its name "because starting from it as a 'place' we easily find arguments."[18] It served various tasks of augmentation, "dilating" (*dilatandi*) and developing arguments from a brief statement, such as a maxim or a law. The method was widely taught in the late Roman empire, which accounts for its

16 *Topica* 71: "nulla praeterea sedes argumenti quaerenda sit."

17 I have much more to say about these matters in the chapters that follow. On the honorable place of enthymeme (or "rhetorical syllogism") in Aristotelian logic, see Burnyeat 1994 (1996). Cicero also defines (in *Topica* 30) the logical category of *genus* as a complete whole divisible into the fixed number of parts or *species* that constitute it; each particular species (*forma*), however, using a logic of analogy (*similitudo*), can then generate an indefinite number of "relations," like a growing family stemming from a shared progenitor. The metaphors of generation used for discussing how minds learn and create are ancient and plentiful, as is the nurturing relationship of *mathesis* (number, including geometry) and language.

18 Kennedy 2003, pp. 42–5, at 43. See also Kennedy 1999; Murphy 2001, pp. 35–143. Augustine's discussions of *memoria* in *Confessiones* X and *De Trinitate* IX–XI show the profound influence on his anthropology of such "topical" analysis.

familiarity to so many late antique thinkers, including Augustine and Jerome; it is basic also to methods of medieval biblical commentary and sermon.

The potentially persuasive proofs themselves – the *copia* as Erasmus called them – were said to reside in *loci* (Latin) or *topoi* (Greek), both words meaning "places." To quote rhetoric historian Heinrich Lausberg, "[t]he loci are … search formulas and in their totality *a reservoir of ideas* from which fitting arguments can be selected."[19] By at least the early thirteenth century, the word *locus* was used as a metonym for a logical demonstration, as in the phrases *locus ab auctoritate* ["proof by means of an authority"] and *locus classicus*, the phrase we still use for a standard proof. It is the basis of the mnemonic "Method of Loci" still taught by some psychologists for improving people's memories. Yet, as I have argued before, what is now called an "art of memory" is not for purposes of mental storing *per se* (setting memories by rote habit), but for the purpose of inventorying and so reliably being able to find and connect a great many things quickly within an existing store when one composes new work.[20] This can give to one's rhetorical proofs the socially persuasive character of "neighborliness" – what we now understand when we judge that something is analogous to another thing – not just its binary "same" or "opposite."[21] It is fundamental to understand this about these systems. Each *locus* or *topos* is first of all thought of as an address within a content-archive that is complexly inventoried: hence the common ancient and medieval model of a richly stocked and ordered mind as a library. The "Method of Loci" is a navigational and reasoning path to finding matters that are retained by the mind in several ways (and it is best if you have more than just one path to each) that enable them to be "called back into mind" (*recollectio*) to serve as the materials from which to make one's new compositions.

How and why Hugh understood such a commonplace of the language arts to be also a property of the mathematical subject of Geometry is a question central to how diagrams are useful as rational *instruments*, and not only as condensed transmitters of some body of received knowledge to be rote-learned by beginners, such as subject-focused topographies and anatomies. It requires studying and mastering as one does a *language*. To quote the historian James Franklin:

19 Lausberg 1998, p. 171.

20 Carruthers 2008 (1990). If readers learn nothing else from that book, they should at least understand that essential point.

21 "Neighbourliness" is Sorabji's translation of Aristotle's *syneggys* (*De memoria et reminiscentia* II, 451^{b}, trans. Sorabji 1972). On the history of similar locational methods, see Carruthers 2008.

> Euclid's *Elements* is not a picture book of shapes. The point of Euclid is to *reason* about the diagrams, and expose the necessary interrelations of the spatial parts [italics in original].[22]

Franklin then observes that Galileo's famous saying that "the universe is written in the language of mathematics" (a language which we now understand mainly in algebraic statements) is, in Galileo's words, one whose "characters are triangles, circles, and other geometrical figures, without which it is humanly impossible to understand a single word of [the natural order]." For Galileo, Geometry (not algebra) best described the building blocks of creation, as it did also for medieval writers and continued to do through the seventeenth century, not only for natural creation and the human arts, but for creative human thinking generally.[23] In a brief work on "Lines, Angles, and Shapes," the English scientist-bishop Robert Grosseteste observed some four hundred years before Galileo, "The benefit of considering lines, angles, and shapes is of the utmost, since it is impossible to understand natural philosophy without them."[24] This is a medieval commonplace. And I have discovered, thanks to the digitized manuscript collection of the great Carolingian library of the monastery of St. Gall, that Hugh's linking of Topical Dialectic (what we would call basic logic) with Geometry is not unique after all – an early eleventh-century manuscript of various works "De Topica" includes, situated between Cicero's *Topica* and Boethius's own *De topicis differentiis*, a work on *Geometria* then thought also to be by Boethius, plus another pseudo-Boethian work "On rhetorical cognition" ("De rhetorice cognatione").[25]

22 J. Franklin, 2000, p. 53, citing Galileo, *The Assayer*, in *The Controversy on the Comets of 1618*, trans. S. Drake and C. D. O'Malley (Philadelphia, 1960), p. 184. See the fruitful discussions of contemplative diagrams by Hamburger 2013, Newman 2013, and the essays in Lutz et al. 2014; and, for a different sort of diagram, Even-Ezra 2021, a study of the logical Horizontal Tree diagrams common in the margins of later medieval academic treatises.

23 See especially Albertson 2014, and the essays collected in Burnett 2010. On the algebraicizing of traditional geometry that occurred over the seventeenth century, see Grattan-Guiness 1996.

24 Grosseteste, *Tractatus de fractionibus radiorum* 1 (ed. Baur): "Utilitas considerationis linearum et angulorum et figurarum est maxima, quoniam impossibile est scire naturalem philosophiam sine illis." My translation. This idea was strongly affirmed by Thierry of Chartres and other twelfth-century neoplatonists, its origins rooted in *Timaeus* and the commentaries thereon: see Albertson, esp. pp. 110–18; G. R. Evans 1976; Oosterhoff 2014; Lesieur 2010.

25 St. Gall MS 830, written in Mainz in earlier eleventh century and held in the library of St. Gall since soon thereafter; it is digitized online (www.e-codices.unifr.ch/en/list/one/csg/0830); see also L. Smith et al. *Codices Boethiani*, II. 204–5. The

So the medieval synergy of rational analysis, inventing new compositions, and geometry was not unusual (it seems a mistake to compartmentalize them as wholly discrete skills even now). An ancient Latin etymology found in Varro, and disseminated widely through the Middle Ages via the encyclopedia of Isidore of Seville,[26] derived *mens* (mind, reasoning) from *memini* (remember). Invention involves an orderly, reasoned search through what one knows, gathered up and composed both in one's mind and with one's books, for learning is rationally stored and accessed as in a library, reliably discoverable through patterns of associations mentally laid out within it, and thus made ready for new uses.[27] Latin *ratio* refers to orderly ways of proceeding, one's chosen *via rationalis*, both "plan" and "method," meanings I will continue to explore throughout this book. English *method* itself is, via Latin and French, derived from the Greek word for a path, *hodos*. Rational, conscious thinking starts with an *investigatio reminiscentiva*, a recollecting search for various "matters" (*res*) using the inventories of one's learning located and connected in interacting libraries, both the ones in one's mind and also the ones in one's books marked out via what is now called a book's paratext, the pathways indicated through underlinings and various glosses, figures, comments, and notes, whether between the lines or in the margins. Memory too has its paratextual links and routes – learning to make and use these is the essence of *ars memorativa*. Such associative schemes are always relational, a matter of *where* and *how* to find a particular matter amidst many others. If their books become dislodged (for example when moving house) from their familiar, habitual haunts on shelves or in piles, scholars often experience feelings of loss and confusion because they cannot quickly "find" (the root meaning of *invent*) the matter wanted in order to continue developing a thought, and such frustrating displacements can be altogether fatal to any composing process.[28]

---

pseudo-Boethius geometry treatise is illustrated by a number of figures keyed to its various definitions. St. Gall MS 817, which was written about the same time (first half of the eleventh century), contains texts of Aristotle's *Categories* and *On Interpretation*, plus Boethius's commentaries on them, and also includes a page with geometric figures, many of which – though not all – are quite differently visualized from those in modern textbooks (see Figure 2, and compare those in Figure 18, from St. Gall MS 830, made at a similar time). I discuss these further in Chapters 3 and 4.

26 Isidore of Seville, *Etymol.* XI.13: "Nam et mens memoria est [For indeed mind is memory]"; ed. Lindsay, my translation.

27 Carruthers 1998 and 2008; also 1996 (on the library model) and 2010 (on pathways, *ductus*).

28 The cognitive thesis of "Extended Mind" offers a cogent account of this phenomenon in modern terms, arguing that humans naturally incorporate the material tools

At the start of the fourteenth century, the Oxford arts master Walter Burley, taught that the only proper action of remembering was just this rational investigation of what we know. "Rote" reciting is not recollection, he says, because rote-retained matter (such as numerical sequences and alphabets, or passages of poetry and prose learned by heart), having been made a habit in us, is thus always present in our minds. With rote, once we start the sequence, the items just roll out in their habitualized sequence, one after another like a wheel (*rota*) that once started just keeps on rolling – that is, after all, the point of learning things by rote. Spontaneous remembering is "improper," Burley says – *abusive*, a term used in Grammar. It is not an ethical or aesthetic value term, though often it has been abusively understood as such (hence Lady Grammar's notorious punitive switch). It is "abusive" only in that it is non-useful (*ab-usus*) as teachable method (*ars*), like that of a speaker who doesn't understand his language's syntactic logic: he may get lucky once in a while, but his grasp of his learning is neither completely reliable for his own uses, nor teachable to others, together with methods they can in turn use to communicate their own thinking and inventing to yet others.[29]

Thinking inventively is also modeled as both hunting and building, activities that are fundamentally relational and locational in nature. The English words "quest" and "question" are both derived, via French, from Latin *quaestio*, "search." And the margins of many books intended for prayerful

---

they commonly use into the mental shaping of their thoughts: "[t]he human mind emerges at the productive interface of brain, body, and social and material world" (Clark 2008, pp. 218–19). The sharp distinction that has been drawn between mental memories and physical books (a variant of an overly binary orality-vs-literacy thesis) thus misses an essential point about how human minds cognitively habituate their familiar tools and physical settings, including their books. See esp. A. Clark (1997 and 2008), J. Sutton (2017), Heyes (2018); see also Carruthers 2008, pp. 309–24, on medieval marginal book decoration as a memory tool.

29 Burley, *Expositio de memoria*: "[T]his properly is recollection, when someone (starting) from a consciously known cognition seeks out the notional trace of a forgotten matter. Sometimes people are not (intentionally) seeking after a particular memory, but in proceeding from an earlier conscious motion toward a later memory, they may happen upon the notion of the forgotten matter, but this is not properly recollection but improper. [Hoc proprie est reminisci quando scilicet aliquis ex intencione inquirit noticiam rei oblite. Aliquando aliqui non querunt propter intentionem sic procedentes ex motu priori in posteriorem deveniunt in notitiam rei oblite sed hoc non est proprie reminiscencia sed abusive."] I am quoting my transcription from London, Lambeth Palace MS 74, a fourteenth-century copy of this digest of teachings *de memoria*, drawn from Aristotle-Avicenna and Averroes (Ibn Rushd); *Expositio* is likely Burley's teaching notes, prepared when he was a *magister* in the *artes* curriculum at Oxford in the first decade of the fourteenth century. I am presently preparing an edition of this unpublished text.

meditation, such as psalters, are decorated with scenes of hunting to invite their users to "keep in mind" this purpose of their reading. Building is all about geometry, whether or not an early medieval mason or abbot had formally studied Euclid.[30] Later scholastic philosophy added another Latin etymology to Isidore's, deriving *mens* also from *mensura*, thus emphasizing the intimate link between reasoning and "measuring" with the shapes you have selected as the tools of your thinking.[31] Thomas Aquinas says that, by metonymy "nomen mentis a mensurando est sumptum [the noun *mens* is taken from *measuring*]" and from *metiendo*, which also means "measuring."[32] Cardinal Nicolas Cusanus, the author of the imaginary bowling game with which I began this chapter, wrote that "mind is that by which we know the limit and measure of all things. Indeed I think that mind (*mens*) is derived from the action of measuring (*mensurando*)."[33] For Aquinas as for Cusanus, measuring, like recollecting locationally as well as temporally, is just what a human mind *does*, the basis of all its conceptualizing and reasoning abilities.

The Conditions of Invention: Solitude, Dislocation, Disquiet, and Perplexity

At the beginning of his major treatise on "The Mind's Way toward God" (*Itinerarium mentis in Deum*), Saint Bonaventure describes how such a mental journey begins, using as his model St. Francis's visionary encounter with a six-winged seraph in whose bosom also appeared an image of Christ crucified. This is how Bonaventure describes his own experience:

30 Binski 2019 and 2012; also Kidson 1990, 2003; Bork 2011; S. Murray 2014.

31 Certainly the similarity of *mens* and *mensura* was a trope extensively played upon in Latin meditations on human reasoning, notably by Augustine (as in *De Gen. ad litt.* IV.4), but also others, including Hildegard of Bingen (see esp. *Liber divinorum operum*: Bk. 1, visio 4, 53–6). But it doesn't seem to be claimed as a full *etymologia* before the thirteenth century. The Latin puns are not possible in Greek, since *metros* ("measure") and *nous, noein* ("reason"), *psuche* ("soul") and *mneme* ("memory") neither sound nor look alike.

32 Thomas Aquinas, "Quaestiones disputatae de ueritate," Q 10, a 1, resp., and also his "Lectura romana in primum Sententiarum," dist. 3, art. 3.3.1 "quid sit mens."

33 Nicolas Cusanus, *Idiota de mente*, Bk. III.1.57: "mentem esse, ex qua omnium rerum terminus et mensura. Mentem quidem a mensurando dici conicio." This dialogue, composed around 1450, is between an academic "Philosophus" and a layman, the "Idiota" of the title, someone without a university education. The layman proves to be a sharper participant than the Philosopher, a convention that becomes more common during the fifteenth century: see Oosterhoff 2014, also Ong 1958.

> Around the thirty-third anniversary of the Saint's death, by God's will, it happened that I went away to Mount La Verna, as to a quiet place; in my distressed state seeking peace of spirit, and staying there, while I was composing in my mind various mindful ways of ascending to God, among the rest there occurred [in my thoughts] that miracle which in the very place had happened to blessed Francis, that is the vision of a winged Seraph with the likeness of the Crucified. In considering which, it immediately seemed to me, that an envisioning [of a Seraph] could show the suspending of this father['s rational mind] while in ecstatic contemplation, and the way through which one arrives at it.[34]

Bonaventure, in some turmoil of mind occasioned by the anniversary of Francis's death, seeks out a quiet place, to be alone.[35] He chooses a mountain that is not at all on his habitual daily routes. That it is also the very spot of Francis's miraculous vision is of course integral to his task, but it is his dislocation from his ordinary habits, his isolation as well as his anxiety of mind and spirit that is emphasized in this account. He is "*anhelo spiritu*," literally "distressed in [his] breathing," and seeking peace of spirit, for *spiritus* should be understood here both physically as "breath" and mentally as his "spirits," his feelings.[36] He goes to that solitary place in order to remember mindfully, as we now would say. Bonaventure speaks of his mind rising up in stages ("*aliquas mentales ascensiones*") from its ordinary thinking process, to turn

34 Bonaventura, *Itinerarium*, Prol. par. 2: "Cum igitur exemplo beatissimi patris Francisci hanc pacem anhelo spiritu quaererem ego peccator qui loco ipsius patris beatissimi post eius transitum septimus in generali fratrum ministerio per omnia indignus succedo. Contigit ut nutu divino circa beati ipsius transitum anno trigesimo tertio [=1259] ad montem Alvernae tanquam ad locum quietum amore quaerendi pacem spiritus declinarem ibi que existens dum mente tractarem aliquas mentales ascensiones in deum inter alia occurrit illud miraculum quod in praedicto loco contingit ipsi beato Francisco de visione scilicet seraph alati ad instar crucifixi. In cuius consideratione statim visum est mihi quod visio illa praetenderet ipsius patris suspensionem in contemplando et viam per quam pervenitur ad eam." (ed. Quaracchi, vol. 8: online at http://www.franciscanos.net/document/itinerl.htm; accessed 24 April 2018); trans. Cousins 1978. My discussion of Bonaventure's encounter was also published in Carruthers 2023, and I use it here with permission.

35 The fact that Bonaventure, as Minister General, was at this time under great pressure from within the Franciscan order, at that time roiled by major controversies, certainly accounts for his mental turmoil, but it is significant that he knows how to turn his anxiety to productive use, by searching out the spirit of Francis in this solitary place.

36 See Chenu 1957 and Jacquart 2020 on the meanings of *spiritus*; on Cistercian spiritual psychology especially, see McGinn 1972, 1977, and the treatise "De spiritu et anima," translated (along with Isaac of Stella) in McGinn 1977.

instead to the task of inventive meditation, "*dum mente tractarem.*" *Tractare* is the common verb in medieval Latin for Invention, the ordinary process of what we now call "getting our ideas." It includes as well the procedures of what we now call "composition," forming simultaneously as we invent them a structure within which we can organize our ideas and proofs (=*argumenta*) as we find them (=*invenire*)[37] and draw them into their places in the composition (Latin *tractare*), along with other associated ones that may follow on in our thinking, either as "equivalent" (*homoiou*), "opposite" (*enantiou*), or "neighborly" (*syneggus*), and which we gather together into an association of "topics" (*collectio* is a frequently used synonym for this process).[38]

Bonaventure is tracting a composition about contemplation, how his human mind can ascend toward God using a structured itinerary for his thinking. It is significant also that he does not speak of intellect in this passage but of mind, *mens*, the type of perception-based cognition rationally derived from our human sensory experiences.[39] There is no suggestion here of miracle for him. Francis's divine revelation was a "miracle" (*miraculum*), but Bonaventure claims no such experience for himself and promises no Pauline *raptus* to his audiences. He offers a journey in distinct stages, an *itinerary* of his mind toward God, not a direct seizure into the divine. Paul, after all, says there is no language in which to tell others of such a rapture (II Cor. 12: 2–4), but Bonaventure is not at a loss

37 Augustine gives the standard etymology of *invenire*: "inuenire est in id uenire quod quaeritur [*inuenire* is to come upon that which is sought]," in *De trin.* X.7. It is always linked to the notion of mental searching – invention does not just happen out of the blue, for one can call out "eureka" only after much questing, as Archimedes's famous cry acknowledges.

38 On these basic logical relationships, see Aristotle, *De memoria et reminiscentia* II, 20 (451^b). In Roman rhetoric, *compositio* is particularly associated with *cursus*, composing one's words sonorously and rhythmically to give one's text a sort of music, and to please and move one's knowing listeners by the verbal cadences. The textbook account is Quintilian, *Instituto oratoriae* IX.iv, but the practice is universal in well-crafted classical and medieval Latin prose, with Cicero's orations as prime exemplars. St. Jerome's account (Epistle 22.30) of his dream that he was denied heaven because he was more attached to Cicero than to Christ speaks to his mastery of such technical *compositio*, evident throughout his prose. A good account of how the scholastic lexicon of tracting translated the earlier, more Ciceronian terminology is Simpson 1986.

39 This psychology is discussed at much greater length in Chapter 2, below. Those using Cousins's translation should note that he sometimes translates Latin *mens* as "soul" and sometimes "mind" depending on context – the Latin just uses *mens* (see the Introduction by Cousins to his translations of Bonaventure, esp. p. 21).The distinction of "soul" from "mind" is much sharper in modern psychology than it was in medieval philosophy.

for language. What Bonaventure does discover (invent) by recalling Francis's experience is both the materials and a dispositive organization for his own composition concerning the stages of contemplation, an *itinerarium mentis* that can lead a human mind upward. Others also can *use* the forms in Francis's vision (the six feathered wings, centrally located about the likeness of Christ Crucified) as a basic compositional outline in their subsequent meditations, seeking understanding of the divine (*in Deum*) through the resources of their own human *mens*. Bonaventure translates Francis's singular miracle into an instrument for shared community use.[40]

Bonaventure was deeply indebted to traditions of monastic meditation that were revivified and adapted by the Victorines and other twelfth-century contemplative orders, the Carthusians, Augustinians, and Cistercians in particular. They all shared a common complex practice, an art or *techne* in Aristotle's term, for reading, meditation, and contemplation – what in classical rhetoric is called *ars inveniendi*, an art of Invention and Composition. Common monastic terms for it are *sacra* or *pia memoria*, translating a Greek term, *mneme theou*, "memory of God." Most often they just called the exercises *memoria*. Jesus's withdrawals into the wilderness to pray in solitude are also clearly within these traditions of preparing for envisioned contemplation, exercises owing both to Hebrew traditions of prophetic meditation and prayer and to Greco-Roman traditions of philosophical invention.[41]

Solitary locations and mental disquiet, including cultivating vigorous, burning (*ardens*) emotions of passionate anxiety and fear sometimes expressed in fierce anger, are part of these preparation procedures.[42] This feature was not

40 The seraph figure soon became a well-used tool for making sermons and other compositions, disseminated particularly by the Franciscans: Carruthers 2009. Bolzoni 2004 discusses sermons preached by the Franciscan Bernardino da Siena that use this same figure.

41 The scholarly literature here is truly vast, and of high quality, including Scholem 1974 (1941); Rabbow 1954; Leclercq 1982; McGinn 1991; E. Clark 1992; Swartz 1992; Hadot 1995; Rapp 2006. The relevant articles in *DS* (available online) are helpful starting points; see also the bibliography in Carruthers 1998.

42 The Latin word *anxius* is often used for this state, though what is meant is different from modern "anxious." For us, *anxiety* is a diseased mental state that needs medical intervention. But inventive anxiety is expectant, an aspect of hope, signaling the advent of a heightened, often visionary, attentiveness of mind and feeling which resolves one's previous disquietude creatively (we now would likely attribute this to an "adrenaline rush," but humoral biology explained it as a balancing rush of warm, wet blood). But unless one can move into a creatively energized state, in medieval as in modern times, one remains merely stuck in a spiritual mire, the monastic state of *acedia*, extreme lassitude and a cold, dry condition called

initiated by desert contemplatives, but in fact is characteristic of Roman rhetoric more generally. In Book X of his *Institutio oratoria*, Quintilian describes those in the throes of inventive meditation as "lying back looking at the ceiling, mumbling to [them]selves to stimulate thought."[43] They also walk about in their bed-chambers, making "those gestures that accompany our stronger feelings, and themselves help to stimulate thought – waving the hand, contorting the face, goading and sometimes striking ourselves" or the table, biting our nails, and often talking aloud with ourselves. This is, he says, the method (the word he uses is *ratio*) that he recommends, along with, of course, continuous practicing (*exercitatio*).

At this stage also, he advises, it is best not to work with a secretary, not only because dictation will slow down your thoughts as you wait for your amanuensis to write down what you say, but also because the intervening presence of another may inhibit the strong gestures and passionate subvocalized inner conversation that stimulates your composing. He also recommends working in a small, familiar room (such as a bed-chamber), and in solitude. Demosthenes "used to hide away in a place where no sound could be heard and no prospect seen, for fear that his eye might force his mind to wander." And so, Quintilian continues, "let the silence of night, an enclosed chamber [*clusum cubiculum*] and a single lamp keep us sharp," for "working by lamplight ... gives the best kind of privacy." Latin even has a word for it: *lucubratio*, "mental work done at night by lamp-light," a word used regularly in medieval meditation practice too. At the end of the fourteenth century, the same advice is still given, to "escape from the presence and company of everyone, and to be alone, lest he should be hindered" when praying inventively in meditation.[44]

"spiritual dryness." A somewhat similar spiritual process draws an excessively hot, angry state of *vehemens*, "fierce, empassioned," into the path (*tractare* again) of creative composing. The Archpoet characterizes his initiating mood in *Carmina Burana* 191 in this manner: "aestuans intrinsecus ira vehementi/ in amaritudine loquor meae menti" (see Ziolkowski 2003 on the passion of *amaritudo*). Cf. also Richard of St. Victor's *Tractatus de quatuor gradibus violentae charitatis*, which begins "I am wounded by love" ("Vulnerata charitate ego sum"; PL 196.1207C). *Vehemens* is a favorite word in Richard's contemplative method.

43 Quintilian, *Instituto oratoriae* X. 3. 15–27; ed. M Winterbottom, and the basis for the revised, 2001 LCL translation by D. H. Russell. All further citations are to this edition and translation.

44 Walter Hilton, *Scale of Perfection* I.30 (English "scale" meant "ladder" and "steps" at this time). Hilton (d. 1396) was an Augustinian canon, writing in English a guide for contemplative prayer, which is anchored in the Victorine way of meditation that is the subject of my next chapter.

The twelfth-century Benedictine abbot Peter of Celle characterizes his chamber (*cella*) as the antithesis of the burdens of his regular duties, a room of silence in which "I draw deeply from the quiet which has now been granted me. The mind has a more extensive and expansive leisure within the six surfaces of a room than it could gain outside. In fact, the smaller the place the more extended the mind, for when the body is constrained the mind takes flight."[45] This play of leisure (*otium*) against daily busy-ness (*negotium*), expressed in the inverse relationship of physical constraint and mental expansion is a common theme in twelfth-century writing on meditation. Contemplative reading (*lectio divina*) was known also as *otium negotiosum*, "busy leisure." The pagan philosophers, Peter says elsewhere, "sought quiet, hidden places, where they could cultivate the mind," and in this respect, he says, the claustral and philosophical disciplines "embrace in each other's arms and kiss away from the crowds."[46] The smallest and most constrained place within the *cubiculum*, or bed-chamber, was the bed (*lectulus*) itself. Undoubtedly impelled by its role in the *Song of Songs*, the meditational favorite of some memorable twelfth-century composing, the bedroom is the innermost sanctuary, in which leisure works the hardest. This can be done in the daytime, Peter says, when minds "removed from earthly activities … graze among the lilies" of their reading. But it is even better when done in "the silence of the middle of the night."[47] And, paradoxically, such constrained enclosure liberates his meditating mind. This too is a pervasive theme in monastic rhetoric.

Picturing Invention

The conventional phases of and instruments for composing are depicted not only in verbal accounts but in pictures as well. Creating any composition was said to start off with an act of "picturing," envisioning the possible shapes of one's composition in some detail before setting pen to paper, or stylus to tablet. "The mind conceives and reason nurtures what the act gives birth to" as Alan of Lille succinctly sums up this process.[48] Craftsmen in other materials counsel (and practice) similar sketching and pre-drawing, in sand or on

45 Peter of Celle, *On affliction and reading*, trans. Feiss, p. 139. The essential study of Peter's spiritual works, and indeed of twelfth-century spirituality generally, is Leclercq 1946, which contains the first published edition of *On affliction and reading*. Leclercq 1963 studies *otia monastica* in particular. See also Carruthers 2018.

46 Peter of Celle, *School of the cloister*, 8.2, trans. Feiss, p. 86.

47 *Ibid*, 15.1; trans. Feiss, 93–4.

48 *Anticlaud.* III.352: "mens gignit, nutrit ratio quod parturit actus." Note again the (ubiquitous) conception–gestation–birth model for conceptualizing and knowing.

another ephemeral surface; vellum surfaces were pre-lined before the scribes began to write. The figure of *Geometria* in Alan of Lille's *Anticlaudianus* is clothed, like her six sisters (the other Liberal Arts), in a fine robe. Hers glows with a great many linear forms (*gramme*), but it is also scattered all over with a fine dust, a detail taken from Martianus Capella, whose *Geometria* carries her *abacus*, the sanded (thus reusable and correctable) wooden board on which people learned and worked with geometric figures.[49] The emphasis in this action is on *planning* carefully the forms one wants to make; that is what the *pictura* that each Art is described as making accomplishes. Geoffrey of Vinsauf likens this pre-planning to the mental plans builders and carpenters "see" for their structures, and to maps that plan the itineraries of a composition's journey, what is called in rhetoric the *via* and *ductus* of the work.[50] All verbal compositions have a planned *ductus* (or should have), also paintings (where the "road" to be traveled through the painting is often depicted within it), and buildings, and of course music, which is said to "go" as one sound leads on to the next in its sequence. Nearly three centuries later, a set of poems and paintings called *Les douze dames de rhétorique*, which shows the stages of inventing and fashioning as twelve handmaidens to Lady Rhetoric, depicts "Deduccion louable [praiseworthy reasoning]," the penultimate stage, as a maiden with her finger pressed to her head as she inwardly envisions her finished work (Figure 4). She has her square rule, compasses, and plumb line in her hand, her mallet and axe within reach, and many half-finished timbers, the "lumber" [=*silva*] with which to make her composition, ready for final assembly and finish. On her table is a partly written book, atop the "precious gems" and "flowers" of her learning laid out in orderly fashion for her recollective investigation to find securely, and just beyond her solitary cell rise two unfinished buildings that she is also making.[51]

In Book I of *Anticlaudianus*, Alan of Lille describes Nature's great hall, where she convenes various powers to envision the education of a liberally educated New Man she wishes to create in order to correct the inadequacies (including too much attention to Aristotle) of the old model, by revising and

49 *Anticlaud.* III.480–83; compare *De nuptiis*, VI.575, where Geometry's abacus is described as being "decentem quandam atque hyalini pulveris respersione coloratam velut mensulam [like a small ornamented tablet colored over with a scattering of glassy-green dust]"; my translation.

50 Geoffrey of Vinsauf, *Poetria nova* I, lines 43–70 especially, but all of section I is relevant.

51 The work has been edited, together with its pictures, by Cowling 2002, who gives a full account of its complex and playful courtly authors and audiences. I discussed rhetorical composition, pre-visualizing and *ductus* (including this painting) in Carruthers 2010, pp. 190–213.

shaping it up with "a finer file," *limax melior* (I. 23), the conventional correcting and finishing file that all imperfect human works need. These powers are summoned in order to "quiet her anxious mind" and help her realize her human improvement plans by summoning up the Liberal Arts. Nature's palace is filled with paintings, faithful images of each of her creations, so real that each can scarcely be told from the actual thing. In this way, Alanus writes, painting (*pictura*) brings what does not exist into existence, "O the exceptional miracle of painting! It brings to being what has no power to actually be, for painting, an ape of truth, sporting with rare skill, turns the shadows of things into things themselves."[52]

One should never just plunge haphazardly into inventing. Places, locations, and their relationships are all encompassed in the term *ratio*, the authorial plan or *consilium* as it is also called, that enables intelligent composing. In painting, this is what a *pictura* was considered to be, for a "picture" is an *organization* of many images. *Mens*, *ratio*, and *imaginatio* work together from the start, not in a modularly sequestered sequence of actions, one beginning only after the previous one is complete.

Even divinely instigated compositions follow the stages from the initial plans shaped from various matters that have been recalled into their places in an envisioned picture, then to the drafting stages made with the help of a stylus on tablets or another surface that is easily erased and changed, and finally to the fully formed, correct "fair copy" on parchment (or built in timbers or placed in mosaics or murals or carved in stone) or recorded in musical notation, as works ready for public presentation and repeated use.[53] The account of making the Kildare gospel book, so wondrous in its painted drawings that Giraldus Cambrensis judges it must have been the work of angels not of humans, describes the stages of composition from initial envisioning to parchment artifact:

> On the night before the day on which the scribe was to begin the book, an angel stood before him in his sleep and showed him a drawing made on a tablet which he carried in his hand."[54]

[52] *Anticlaud.* I.122–4: "O nova picturae miracula! Transit ad esse/ quod nihil esse potest simia veri,/ arte nova ludens, in res umbracula rerum/ vertit." As he also says: "Haec probat, ista facit [That art (=logic) gives proofs, this one (=painting) makes]" (I. 128). On the essential role of play (*ludus*) in all creation, see Carruthers 2013, chs. 1–2. It is not irrelevant that there was a strong vogue at this time for *trompe l'oeil* painting, especially on walls and columns.

[53] Carruthers 2008 (1990).

[54] Gerald of Wales, *The History and Topography of Ireland*, trans. O'Meara, 84–5 (Penguin 1982).

The angel asks the human scribe if he can copy this drawing, but the scribe, with refreshing honesty, responds that he cannot. So, after further divine intervention through prayers from St. Brigit, the angel shows up again on the following night with many more drawings, all those to go into the finished, fully colored book. The scribe-artist takes careful note of them all, committing them to his memory in full ("*memoriae fideliter commendans*") and copies them precisely ("*ad unguem* [to the toenail]") into their proper places ("*loci competentibus… impressit*") in the book. "And so with the angel indicating the designs, Brigit praying, and the scribe imitating, that book was completed."[55]

Figure 5 depicts the famous mystic, Hildegard of Bingen, composing her prophetic visions. It is easy to forget that prophecy is a type of literature that, like all literary genres, has certain formal characteristics that ratify the visionary's divine purposes and visionary experiences to audiences (such as Hildegard's) in a manner that persuades them, in large measure by fulfilling their conventional expectations. The audience members (or *testes*, "jurors," in the traditional analysis of any rhetorical occasion) are convinced it is true prophecy exactly *because* it satisfies these conventions. This image is from the Rupertsberg manuscript of the visions she called *Scivias*; it was made during Hildegard's lifetime, and for a long time was in the library at Wiesbaden, but has been lost since World War II. Fortunately, it was photographed and fully copied before the war, and is now online as the Rupertsberg Codex.[56] Note that the prophet's eyes are closed, for she sees inwardly, envisioning. Similar images of her composing occur in a corner of each of the ten paintings of visions that make up her later *Liber divinorum operum*.[57] The opening page

55 *Topographia Hiberniae* II. 71–72 (in *Giraldi Cambrensis Opera*, vol. 8, ed. Dimock): "[N]octe sequente iterum affuit angelus, eandem figuram aliasque multas ei praesentans. Quas omnes, diuina opitulante gratia, statim aduertens et memoriae fideliter commendans, libro suo locis competentibus ad unguem scriptor impressit. Sic igitur angelo praesentante, Brigida orante, scriptore imitante, liber est ille conscriptus."

56 Many online databases now offer images of this famous picture, the first illumination in the Rupertsberg manuscript. A helpful link is https://commons.wikimedia.org/wiki/File:Hildegard_von_Bingen.jpg. Other copies of *Scivias* exist but without illustrations, a fact which is telling. Evidently its readers were intended to imagine from the words, and envision them with their inner eyes (*ante oculis ponenda*) as they contemplated the texts, an emphasis on the meritorious practice of carefully imagining from verbal *descriptiones*, painting with their minds not simply looking at finished objects. Kessler 2000 is helpful in this matter.

57 There is only the one illustrated manuscript of this work, the so-called Lucca manuscript, made in the early thirteenth century some forty years after Hildegard's death. See Campbell's Introduction to his FC translation, pp. xv, 21–5. Campbell's

of the only painted manuscript of this work, tucked into its lower margin (Figure 6), depicts Hildegard, her mind enflamed directly from the divine being above her, tablets and stylus in hand. There are variations from the Rupertsberg picture. Notably, Hildegard's eyes are wide open as the divine power touches her mind and she waits, tablets in hand. Her secretary Volmer is entirely outside her chamber, bifolium and stylus at the ready, but Richardus, her sister-companion (whose eyes are closed), is in the chamber with her as a faithful witness to the visionary event, even though it occurs beyond her human understanding. Hers is like the role of Alypius in Augustine's visionary conversion to celibacy, discussed below.

In both manuscript pictures notice that Hildegard is composing alone in a small enclosed chamber using a pair of waxed tablets to record her visions, the traditional stages of medieval envisioning-inventing processes. Her human amanuensis (Volmer) holds a closed codex; he is ready to write from whatever text she will give him on her tablets but he has no stylus with him (that will come later in the process), and has placed only his open, witnessing eyes and – prominently – his right ear inside the room. Note that he is *listening*. Hildegard is the only one directly *seeing*, though inwardly with closed eyes; the waxed tablets she holds may well depict the traditional metaphor of memory as the "waxed tablet" of her recording mind. She is illumined by divine light which touches her eyes (sight), one flame reaching also to her ear (hearing) and another reaching down to her neck to where the *spiritus* that animates her pulse and heart is closest to her skin (touch). The conception is spiritual and somatic, body and mind together.

Hildegard's mind and senses are divinely enflamed, yet in that same light *she will write on wax*. That is an important detail. Wax is a human medium, fungible, erasable, correctable, and inadequate to truly express the Divinity she sees when in full rapture – like St. Paul, what she sees is not utterable in mortal languages. And she is not speaking in these pictures. The human words she forms on her wax tablets, both mentally and materially, are necessarily a recollected, insufficient approximation of what is actually inexpressible. The ephemeral medium of wax indicates both the inadequacy of her human words and their changeable nature. That is well reflected also by the different versions of her seeings that are painted in various manuscripts, even those made in her lifetime.[58] These pictures of Hildegard are also a guide to the stages of

translation reproduces each painting in black and white; the color images have been digitized together with the complete manuscript, and are accessible online through the World Digital Library (see link at Figure 6).

58 Caviness 1998; Newman 2005. Manuscripts and murals made from Bonaventure's pictural meditation, *Arbor vitae*, show similar variety in their renderings of the text's instruction to "first, make a tree in your mind"; see Sandler 2020.

composing that are described, for example, in Eadmer's "Life" of St. Anselm, as well as by Quintilian.[59] Hildegard will write her *remembered* visions using human language (likely in several drafts and in mixtures of German and Latin) onto material waxed tablets.[60] Drafts at this stage are called a composition's *res* – we would call them drafts. And evidently one would have more than one such draft, erased and redrafted, augmented and rearranged.

The other arts have their equivalent stages. A revealing example is a sketch by Raphael, now in the Ashmolean Museum (Figure 7). On the same sheet of paper are two kinds of drafts: pencil drawings of human torsos in two positions, one slightly overlapping the other, one clothed and one not, and an outlined human head; and a much-corrected draft of a sonnet, partly written over one of the sketches. To the right-hand side of the draft Raphael has written out a "public" version of the first stanza in a formal, legible book hand. Evidently Raphael regarded all his sketches, whether graphic or verbal, as unrefined and ephemeral stages of composing (neither he nor his contemporaries kept such rough studies, any more than poets kept their "foul papers," and the few we still have survive by accident).[61]

A draft on tablets or (later) onto scratch-paper, when ready to be copied out by a trusted secretary like Volmer, was called the *dictamen* – this fuller version might well then be copied out into a foliated booklet, such as the one that Volmer is clutching. An author might well correct dictamenal drafts as well, adding, subtracting, rearranging, just as we do now, before the work is sent

59 Anselm's trials while composing the drafts of his *Monologion*, involving both scribes and ruined waxed tablets, are discussed in detail in chapter 6, "Memory and Authority," in Carruthers 2008 (1990).

60 It has been taken as evidence for their general "illiteracy" that women visionaries often asked a cleric to correct their initial compositions, an explanation that assumes a vast gulf between Latin and the vernaculars in use amongst nuns and monks alike. I am not convinced of this. It seems to me that stressing an author's use of a particular vernacular underscores the general inadequacy of all human language (including Latin) to truly articulate enraptured revelations. Latin is just the received medium for discourse to be made public, an indication of the texts' status as being final and of record. As far as we know, Hildegard and likely others checked the final Latin versions before authorizing her works to be finally copied out in a formal scribal hand suitable for circulation to others, and it is a medieval commonplace that authors, upon releasing a work to the public, ask their readers to correct their infelicities and errors – this trope even has a name, *limax correctionis*, "the file of correction."

61 Whistler et al. 2017, catalogue #50, fig. 90. See also catalogue #47, reproducing a similar page by Raphael, now in the Albertina (Vienna), that also contains figural sketches and drafts of a sonnet; both sonnets are transcribed and discussed separately in this catalogue, p. 249.

into the world in its "copy text" version, ready for publication in the form of "fair" copies drawn and written on lined vellum leaves in a formal book hand. There might well be several exchanges between author and amanuenses before the author was satisfied (this seems to have been the case for Cicero even with his ever-faithful Tyro, and also for Anselm with his less careful secretaries, not to overlook Thomas Aquinas in his cell with a foursome of scribes, all scribbling down the master's words).[62] The copy text is the authorized text, the text of record which the author is at last content to have presented in public and also held in a library for others to read and to copy. But notoriously, some authors would continue to edit and rewrite their texts, so that for some works there never was a single copy text: the three (or is it four?) versions of *Piers Plowman* come to mind, as do the incomplete redactions that circulated of many vernacular and scholarly texts alike. This textual phenomenon has its own names in literary scholarship, *mouvance* and *variance*.[63]

Probably the most depicted medieval scene of a person composing is the opening of Boethius's *Consolation of Philosophy*. Figure 8 reproduces the frontispiece to the text which was drawn in an English manuscript of the earlier twelfth century. Boethius is alone in his prison cell, a lined bifolium in hand, although his attention is evidently caught by something else. Presumably the bifolium represents the bitter poem of complaint he tells us he was composing with the aid of the Muses, shown blowing their trumpets beside his compositional *lectulus*. A very tall Lady Philosophy is just on the other side of his cell, about to enter. Her instructing finger is poised and ready. In a few sentences more, she will enter, banish the Muses, and begin her teaching to release Boethius from the thick chains about his neck which, in his despair, he has forged for himself (Book I, 2.25–6).

A picture made a century or so earlier in Salzburg (Figure 9) was also a frontispiece for the text. This one shows several scenes enclosed in an architectural frame. In the right enclosure is an unfinished sketch of a childish, tearful, tousled Boethius. Only his head has been drawn, but his youth and

[62] See the "Introduction" to Carruthers 2008 (1990). On the systems of abbreviation used, see OCD, s.v. *tachygraphy*. The entry importantly observes that such figures were used as "a system of abbreviations rather than as the kind of shorthand that would have enabled a secretary to take down a speech verbatim."

[63] There is now a vast literature about this, first analyzed positively for Old and Middle French literature, as in Zumthor 1972 and 1984, and B. Cerquiglini 1989. The essays in Johnston and Van Dussen 2015 provide a good overview. The shift was in large measure a reaction to a previous editorial doctrine which sought out the original author's "intended" readings, presumed to have been obscured by the stupidities and misreadings of later "ignorant" medieval copyists.

troubled mental state are plain. Below him, three disgruntled Muses have been shoved out of a door which is now shut behind them. In the top center compartment Lady Philosophy stands as she is described in the text, a book in one hand, a rod in the other, and embroidered on her robe a ladder with Greek Π at the bottom rung and Greek Θ at the top (for *practica* and *theoretica*). In the bottom chamber, she is at the foot of Boethius's *lectulus*, speaking to him and using her gesticulating fingers to remind herself of her inventory Places, while he (now plainly far more mature and wiser) notes her words on a bifolium, a Tyro to her Cicero, a Volmer to her Hildegard.

Figure 10 shows a similar, though later, scene of Boethius learning and composing. This painting is in a grand presentation manuscript of the *Consolation*, made in Bruges in 1470, containing its first translation into Dutch, by Jacob Vilt, a goldsmith. This manuscript has frontispieces for three of the books. This picture introduces Book V, the most theoretical and appropriately disputatious of the books. Notice that Boethius has left his bed (though he is still alone in his enclosed chamber); his waxed tablets are hung on the wall next to his bed. Lady Philosophy is speaking, indicating her chief topics with her right finger pointing to their positions on her left hand, while Boethius is questioning her with his left finger tapping the appropriate joints on his right hand. This topical, finger-based method for learning and recollection was familiar from various hand diagrams used in learning and meditation, such as the Guidonian music hand and the Hands of Meditation that one finds in many books for study and prayer. Especially in books for laity, "hands" are drawn for the exercise of Examination of Conscience that precedes Confession.[64] Lady Philosophy is still teaching Boethius her main points (literally) using her finger joints as her *sedes argumentorum*, but he has learned enough to dispute with her using the "places" of his own hand (as indeed is the case in the later books of the *Consolation*, when Boethius finally is wise enough to ask intelligent questions instead of just complaining). The set of waxed tablets hanging on the wall is full of writing (presumably his *dictamen* of

64 On hand diagrams see Sherman 2000, Berger 2002. An interesting case is a "Hands of Meditation" drawing showing the various topical "places" on each hand, ready for use but unlabeled, that are drawn at the beginning (fol. 4) of a large fifteenth-century book (Bodleian MS Laud Misc. 156), which was made for learned studying of Nicholas of Lyra's *Postillae litterales*, a much-used encyclopedia of glosses on the Bible. The book does not actually contain Lyra's text, just a detailed topical index to it, several diagrams of the biblical *geometricalia* used in its study, and the diagrams of the *Speculum theologiae* popular with preachers for composing their sermons (this manuscript was most likely the work of Franciscans, though it was in the library of St. John's Hospital in Exeter until the Dissolution). I discussed it at length in Carruthers 2012.

a previous section of text) and ready for a scribe to arrive and write out the draft. Boethius, having carefully remembered his exchanges with Philosophy, will correct these as they go along, and then at last the finished text will be inscribed in a fair copy by professional scribes and rubricators; then their first inscription will be carefully corrected by editors within the Scriptorium, and the illustrations painted where indicated, in order to make a grand book like the one on the ledge at the left, ready to be taken off for public display.

"Argumentum fidei": Faith, Witness, and Forensic Arguing

The proven methods for inventing and arranging one's arguments, as described by Quintilian and many other ancient writers, should also recall Augustine's description of himself in the climactic scene of his conversion to a celibate life. As he tells us in *Confessions* Book 8, his process of thinking through his quandary involves a number of the postures and gestures associated with Invention in classical rhetoric.[65] He first describes himself in great distress of mind and heart after his encounter with his newly converted friend Ponticianus. He encloses himself "*in cubiculo nostro*," his mental inner chamber. He is also arguing aloud freely to himself (though his companion Alypius is present): as he says, "My words … said less about the state of my mind than my forehead, cheeks, eyes, colour, and tone of voice." Then he goes out into the private garden of their lodging, as far as he can from the buildings, Alypius following. There they sit, taking along "a book of the apostle," Augustine in anguish of thought, Alypius silent. Note that Alypius's silence is emphasized throughout the whole scene (as is Richardus's in the Lucca portrait of Hildegard, Figure 6). In his turmoil, Augustine tears his hair, he locks his hands over his knees, he talks aloud – and constructs a formal disputation about how his will can be divided as it were into "two wills," and how this idea should be understood. His arguments are marshalled *in utramque partem*; indeed Augustine uses this very language (particularly in sections 10–11) as he sets about constructing what amounts to a logical division into parts using the methods of rhetorical invention and *dispositio* to build his *argumenta in utramque partem*. His motive is forensic: he is accusatory of himself. But, as his mental debate (he calls it a *controversia*, the name of a school rhetorical exercise) grows fainter, he sees in his mind, as though emerging from one part ("*ab ea parte*") of his unfinished argument, the image of "*casta dignitas continentiae*," Continence envisioned as a lady accompanied by a flock of

[65] I discussed this scene at greater length in Carruthers 2018. My quotations are from Chadwick's translation.

exemplary figures, those *exempla* which are another sort of arguments. This complex, active image serves finally to enable Augustine to think through the divisions of his *controversia* to their optimal conclusion. And Alypius, sitting with him, is still silent.

Most modern readers just pass over this entire passage (though it is very long) as though it were a diversion from the main psychological and therapeutic action of Augustine's vision, which in our culture is all that really counts.[66] That formally structured argumentation leading to a reasoned conclusion could be fruitfully rooted in anxious doubt, tears, wild movements, and gestures – this is to us inconceivable. But remember that Augustine says (sect. 8) that his strong gestures and postures were from the start deliberately willed actions: "If I tore my hair, if I struck my forehead, if I intertwined my fingers and clasped my knee, I did that because to do so was my will."[67] Moreover, Alypius's immobile silence fully respects the preference expressed also by Quintilian, that the procedures of inventing be completely uninhibited. I think this is why Augustine so stresses his friend's stillness. It is a mark of friendship, yes, but also evidence that Alypius comprehended, from deep inside this culture, what Augustine was doing, and waited for his thinking process to reach its conclusion, once he was persuaded to a confident (*fides*), decisive action.

After this extended phase of forensic-style controversy, now weeping greatly, Augustine first stands up, leaves the book, walks some distance away and then throws himself down under a fig tree, prostrate. Lying prone is a posture of composing as well as of prayer and humility. In that posture he repeats to himself some familiar texts (a means to stimulate more of his inventive recollection) and then hears a sing-song, child-like voice commanding him "pick up, read" (*tolle, lege*). At that point, he gets up again, goes back to where Alypius is still sitting and takes up the book he had left, opens it, and reads the fated passage from Romans 13. Historians have emphasized the supernatural in Augustine's account here of using the ancient custom of *sortes*, reading one's fate in a book, but I think this makes the scene appear only superstitious. Even in this late Roman culture omens were used commonly as arguments

66 Peter Brown 2000 characterizes this scene as a "therapy of self-examination," p. 175. More recently, Miles Hollingworth 2013 writes of it as an instance of "antiquity's silent respect for madness," filled with "[t]he passion and deluge of this moment" when Augustine "sits under a tree [*sic*] … and converts, of a sudden, to Christianity [*sic*]," p. 198.

67 Augustine, *Confessiones* VIII.8 (20), CCSL 27: "Si uulsi capillum, si percussi frontem, si consertis digitis amplexatus sum genu, quia uolui, feci." On the artful literariness of Augustine's self-description here, so much a product of his time, see Courcelle 1963, pp. 191–7, and Conybeare 2006.

for doing things (or not). As Augustine narrates it, reading this omen gives him the final convincing argument in a reasoning process that he began long before, the "inspired" moment we all still cherish when everything we have been thinking about suddenly comes fully clear, and we are confidently persuaded of its truth.

It is astonishing how detailed Augustine is about every stage of his argument, every location and position, every posture and gesture. He has taken pains to bring it before our eyes, with the vivid illumination – *enargeia* – that was highly praised in rhetoric, from Aristotle to Quintilian, and cultivated carefully in monastic practices, as among the best means of confident persuasion. Greek *enargeia* translates into Latin, Quintilian pointed out, as *evidentia* and *illuminatio*, that vivid engagement with a juror's mind and feelings that persuades him as a caring witness to what happened (*evidentia*) instead of remaining a distanced observer of hearsay. Recognizing in Augustine's behavior the invention methods he and his culture approved and taught understands this scene to be fully rational as well as emotionally and indeed physically charged.

In the later Middle Ages (1336), Petrarch, revisiting this famous scene nearly a thousand years on, describes how he ascended Mt. Ventoux with his younger brother, a companion he carefully chose, as he tells us, after considering many of his friends, whose inadequacies for the task he describes at length.[68] With him Petrarch carried, as always, his copy of Augustine's *Confessions*. Readers tend to overlook the brother's presence as unimportant: after all, he says nothing on the way up the mountain, finds a far more direct, if more difficult, path than the one Petrarch's fictional self in his stubbornness arrogantly selects, and indeed laughs at his older brother's many wrong choices. But Petrarch as author evidently understood Augustine's original account better than we do now. The brother takes the role of Alypius, the all-important witness to the final action, the one whose eyes and ears we can trust. (There are two servants along also, but, as social inferiors, they cannot in law be witnesses.) Augustine's passionate turmoil as he vigorously thinks through his dilemma is mirrored in Petrarch's fictive course up the mountain, self-described as one of rambling and hesitation, retracing paths, not seeing the way. His is not at all a straightforward march to his destination at the mountain's top, which they finally reach at sunset. Through it all his brother, like Alypius (and Richardus), is his silent companion. Only when Petrarch opens his copy of *Confessions* to read aloud the passage fated for him,

[68] F. Petrarca, *Familiares* IV.1; ed. Dotti. I am indebted to the fine analysis of this scene by Southerden 2018. The exemplary story of Augustine's conversion places "examining conscience" firmly in traditional practices of school rhetoric with respect to forensic analysis and argument, including the convincing importance of *enargeia*. Augustine's text was never unknown: see Courcelle 1963.

the act which unites *his* divided will and converts him to his direct path, does his brother speak, to ask for more to be read out.[69] But Petrarch refuses, recognizing both that the passage is fated for him only, and that his brother has fulfilled his role of witness: "Deum testor ipsumque qui aderat [I call on God and him who was there as witness]." Using the verb *testor* leaves no doubt about the brother's role – he was *there*, indeed "intentis auribus stabat [he stood with intent ears]," like the large-eared Volmer who witnesses Hildegard's visions (see Figure 5).[70]

Andrew Louth has emphasized the distinctively social element in Augustine's account of the heavenly vision he shared with his mother, Monica, recounted in *Confessions* Book 9; it is essential in his account in Book 8 as well.[71] This social ingredient is key in all forensic argument, which requires the testimony of witnesses, *testes*. For the nature of Augustine's cause is what was called in antiquity both *causa anceps* and *causa admirabilis*, in Greek *paradoxa* – a "two-headed" and "impossible" case in which both sides (*partes*) are equally true but logically contradictory, and also a cause towards which the jury is hostile and unbelieving.[72] In Pauline Christianity this is the triple impossibility in ordinary forensic debate that is inherent in all *argumentum fidei*, the "arguing by/for faith" as expressed in Hebrews and Romans:

> Now faith is the substance of things to be hoped for, the evidence [argumentum] of things that appear not. For by this the ancients obtained a testimony [testimonium]. (Heb. 11: 1–2)[73]

[69] The text Petrarch reads is a portion of *Confessiones* X.8.

[70] Southerden 2018. Petrarch's letter is a carefully crafted work of literature, not at all spontaneous (as it claims conventionally to be) but likely composed many years after the supposed date of the ascent in 1336.

[71] Louth 1981, pp. 136–8. Augustine gives Monica a major role during his ascent to vision, as he does to Alypius during his conversion process. I am suggesting that this social emphasis arises in part from his immersion in the Roman values that are embedded in rhetorical practices, for in rhetoric, persuasion to confident belief is always, fundamentally, a social activity: it requires participating "judges," not passive spectators. It is also worth remarking how pointedly Augustine emphasizes his psychological and physical torment (from a fierce toothache and hoarse breathlessness upon leaving his teaching position) as a prelude to his account of Monica's healing visit and their subsequent shared vision (Bk. IX.4–6).

[72] The *causae* are defined by Cicero, *De inventione* I.xv.20–21. A third type of case, *obscura*, is also applicable to the Christian *argumentum fidei*, which in this way is a three-fold impossibility in human rational terms.

[73] "Est autem fides sperandarum substantia rerum, argumentum non apparentium. In hac enim testimonium consecuti sunt senes." Note how argument and witness are emphasized in these verses, as they are in the Greek text: "Εστιν δὲ πίστις

Alypius is an eye-witness, his presence in this scene foregrounded. But of course it is exactly what *cannot* be seen that constitutes ultimate belief, *fides*, which the readers whom Augustine is really addressing must be persuaded to hold in full confidence. They are equally witnesses and judges of his arguing.

Paul defines his argument with Greco-Roman philosophy and science specifically as paradox. This is clear in his epistle to Romans:

> For the invisible things of him, from the creation of the world, are clearly seen, being understood by the things that are made; his eternal power also, and divinity.
>
> For we are saved by hope. But hope that is seen, is not hope. For what a man seeth, why doth he hope for? But if we hope for that which we see not, we wait for it with patience.[74]

It was a serious problem for Christian apologists like Ambrose, who wrote the earliest Latin theological meditation on the Six Days of Creation (*Hexaemeron*), to make the case that educated pagans and even recently converted Christians brought up in Greco-Roman cosmological ideas of a changeless Perfect could confidently believe, that a single all-powerful, eternally perfect Being would voluntarily *choose* to express its essential completeness in the manifold formations of a Nature created as a community of parts. It was even more shocking that God would choose to *lower* his own Divine Being and become a mortal human in Incarnation. And of course that the divine Being (*esse*) was Three, not One, was suspected by many Hellenized Romans, even those professing Christianity, to be another way of lessening, because it was partitioning and so diminishing, the Single perfect being.[75] Why would a Perfect Being *choose* to be less so? But this matter anticipates questions that are discussed in the next chapters.

ἐλπιζομένων ὑπόστασις, πραγμάτων ἔλεγχος οὐ βλεπομένων· ἐν ταύτῃ γὰρ ἐμαρτυρήθησαν οἱ πρεσβύτεροι."

74 "Invisibilia enim ipsius, a creatura mundi, per ea quae facta sunt, intellecta, conspiciuntur: sempiterna quoque ejus virtus, et divinitas." (Rm. 1.20). "Spe enim salvi facti sumus. Spes autem, quae videtur, non est spes: nam quod videt quis, quid sperat? Si autem quod non videmus, speramus: per patientiam exspectamus. (Rm. 8: 24–5)

75 Good accounts of some of the issues at this time concerning Creation include Marrou 1958 and Pelikan 1993. See also McGinn 2002 and Albertson 2014.

2

Investigation and Invention

Fantasy, Imagination, and Wonder

Among several detailed, step-by-step medieval guides to contemplative technique that were composed within the century between about 1165 and 1270, two stand out as particularly influential: Richard of St. Victor's *De contemplatione maior* (alternatively called "The Mystical Ark" or "Benjamin Major" in English) and greatly influenced by it, composed around one hundred years later, Bonaventure's *Itinerarium mentis in Deum* ("The mind's journey into God"). In this chapter, I focus on the central importance of "envisioning," a particular kind of mental *visio*, as the primary engine of meditation for theological contemplation. In medieval monastic life envisioning was a learned craft practice, not reserved only to momentary rapture (though for a very few, it might include such a rare, divine gift), nor was it a medical or drug-induced condition (though my own students have sometimes assumed it must have been). Indeed, the craft of mental envisioning was not thought to be different from reasoning, but the first essential step in every sort of rationally imagined composing.

Rhetoric is a profoundly social art, directed primarily to composition with the goal not of "self-expression" nor of communicating over a distance to some statistically constructed market of consumers, but of publicly persuading a particular audience of various people who are physically present with you in a particular place and time to do something, such as to believe you because they have enough trust (*fides*) in what you say to be willing to believe and act as you have convinced them to do. Cicero describes how civil society itself came about when a great and wise man realized that the ignorant, violent, and lawless humans that were wandering about, quarreling like animals, had a potential to become useful and honorable in their occupations, and "through reason and eloquence" persuaded them to act with kindness and dignity through rationally organized civic life.[1]

[1] Cicero, *De inventione* I. ii. 2–3; trans. Hubbell, LCL. This founding myth was much invoked throughout the Middle Ages, as well as Cicero's immediately following

Reflective thinking is an activity of rational investigation, voluntarily directed (*investigatio rememorativa*) among one's conceptualized experiences (*noeta/ notiones/ intentiones*), using as its instruments reason and imagination (*vis rationalis et vis imaginativa*), in order for the intellect to be able to comprehend particular sense-derived notions as generalized ideas that can be understood in terms of truth or falseness. The relative materiality of the mental data (*imagines, phantasmata*) as they process and are processed through the phases of this complex activity is key, for the intellect, being entirely immaterial, can only comprehend and understand (Plato says "recognize") ideas that are universal: whereas human sensory comprehension is based, as its name indicates (*experimentum*, in Greek *empeiros*), wholly in material particulars, i.e. the individual data from the five externally sensing organs that are received materially in a common area (*sensus communis*) of the head.[2] So the mentalizing process of concept forming is fundamentally one of increasing abstraction, de-materializing, and "mentalizing" (or "spiritualizing") particular material data into a form discoverable (*inventibilis*) by an *investigatio rememorativa*. A *notio*, however, is never entirely immaterial – hence the fundamental bridging role of *reminiscentia* when rationally (consciously and voluntarily) directed by an intellectual action. *Ratio* (*vis rationalis*) is an activity serving Intellect, and the construction of various *notiones* into a *via rationalis* is effected during an *investigatio rememorativa* (synonyms being *recollectiva* and *reminiscentiva*).

It is an important linguistic fact that in the languages/lexicons that form the matrix of western psychology and philosophy of Mind (including pre-modern work *de anima*, "on the soul"), there are not separate roots for storing and retrieving memories. *Memini* and *reminisci* are of the same root (and, as we saw in the preceding chapter, so is Latin *mens*). This is generally true of Indo-European and Semitic languages, of Germanic, Greco-Latin, and also Arabic and Hebrew.[3] No language in these groups has wholly different verbal roots for making a memory and for recalling it, only a single "family" (*familia*) of words deriving from one root signifying "past matters

caution that a person with an agreeable manner but only "a depraved imitation of virtue" without any sense of moral duty, will "corrupt cities and undermine the lives of men."

2 For Aristotle, this sensory reception area was near the heart, but other ancient medical traditions including Galen placed it in the head, as does Augustine. Theirs is the dominant medieval tradition. See Carruthers 2008, chapter 2, and the bibliography referenced there.

3 French *se souvenire* linguistically calques Latin *invenire*; French also has a panoply of words deriving from *memor-*, *mémoire* and *remembrer* among them.

recalled in the present." When there is an entirely separate verb for mentally securing what we now call "a memory" (note the rather misleading article required in English),[4] it derives from a root that means "store" and "guard" (as in Greek *thezaurizein* and Latin *vis conservans*, Hebrew *shomer*, Arabic *ḥāfiẓ*), and is unrelated to the lexical root in those languages that means "remember, recollect." In his treatise *On Memory and Recollection*, Aristotle admits that Greek *mneme* commonly meant both "record" (as "store") and "remember." It is his need for clear analysis of the stages of the cognitive process that led him to coin the term *anamnesis*, derived from *mneme*, to distinguish the mental stages of making a perception potentially recollectable (i.e. "memorable"), from the temporally later action of recollecting that experience when what was being perceived *then* is no longer immediately present *now* to your senses.[5] The important questions to consider are not how exact our recollection or mimetic reconstruction is of some completed event (its "accuracy" in a modern sense) but why and for what present and future purposes we are recollecting it now.

Grammatically, you only have a memory of some content when you call up that content *in the present*. You only even know that you actually have a memory of something when you re-present it by remembering it. This is why we have no actual memory of our own particular birth, though we understand rationally that we must have been born.[6] You have a memory in the same way

4 Because of this grammatical difference, forms that in Latin indicate activities, such as gerunds and gerundive adjectives like *imaginativa*, especially when they occur without the noun they are understood to modify (in this case *vis*), when they are translated into English and other article-requiring languages, tend to be understood as simple nouns naming "a person, place, or thing" and thus are nominified as if they were material, space-occupying things in themselves rather than instruments and phases within a single complex activity. Unless one is careful, the English translation can compromise the logic of the Latin.

5 At the start of *De memoria et reminiscentia* II (451^a 19–24). In *De memoria et reminiscentia*, Aristotle is especially at pains to clarify the distinction. That same philosophical need for definitional clarity also drives scholastic analyses of memory and recollection. But in more popular uses, both vernacular and Latin, "memory" and its verbal "family" refer to any actions of recollecting things that are temporally past and gone in the present. Pre-modern linguistic word-lists, from the time of Varro, are organized conceptually into just such clusters (*familiae*) of words: an excellent medieval example is the thirteenth-century *Derivationes* by Bishop Uguccione da Pisa. Modern anthropological linguistics now calls these "radial categories"; see the discussion by Lakoff 1987.

6 A useful discussion of these matters, focused on Duns Scotus but in helpful comparison with other Scholastics, is Cross 2014; also the relevant chapters in Marenbon 1987. As Aristotle commented, we cannot remember being born because at

you have a thought: you must *make* it in the present. The images by which you know your *notiones* ("concepts") must be formed by your *vis imaginativa* and *vis rationalis* working together each time you recollect. Memory functions always *in the present*. The initial image (*phantasm*) formed from all the sense-data received into *sensus communis*, like the particular external sense-data themselves, is no more, it is gone – and our minds cannot know content without some form of it, however schematic, by which to know it. Mental *notiones* must be imagined, and thus called productively into the present, and so "re-presented" in our minds. For medieval people, remembering was never mainly about past events. Memories are for and in the present, created for thinking now and for planning ("fore-seeing") possible futures.[7] Hence the dual meaning of Latin *recordare*, both "recount" and "record," for what we relate as our memory of something is always only our own *account* of an experience now finished (and when recording information that needed to be kept exactly for posterity, like an edict or historical dates, it was always written down).

It is also significant that the Latin verb *memini* itself combines past and present time, for although its meaning is present its forms are those of the perfect (past) tense. The forms of many Latin verbs for the various procedures of thinking are also telling. Many such verbs are deponents (though they may have simple active and passive forms as well).[8] Deponency is a linguistic means adopted in Latin to indicate the "middle voicing" of a verb, a feature of Greek syntax but not native to Latin. It represents an action that combines both active and passive meanings, receiving and acting simultaneously. Latin

that moment we had no way to conceive of the act of our birthing as a specific experience within time that we can now recall; we had, as it were, no "place to connect it" in our mind. We learn from others that we were born, and it is their statements that we remember. Modern psychology somewhat similarly sometimes distinguishes "episodic" from "semantic" memory.

7 See Carruthers 2008; Dudai and Carruthers 2005; Suddendorf and Corballis 2007.

8 Grammatically, *memini, -isse* is classified as a "defective" verb, hovering between past and present without the full formal complement of either. Deponent verbs often have an alternative active form as well, such as *memorare* and late Latin *memorari* alongside *memini* as a possible (though rare) infinitive form of *memor*, both verbs having their full range of possible inflections, unlike *memini*. On these grammatical concerns, which as logical questions occupied the Modistae grammarians of the twelfth through fourteenth centuries, helpful discussions are Rosier-Catach 1983 and 1994, L. Kelly 2002, Lombardi 2007; Bursill-Hall 1971 also contains much helpful material as do the collected papers of R. W. Hunt 1980. On the late classical expansion of Latin deponent verbs, many newly coined, see Flobert 1975.

deponents include *miror, mirari* ("marvel"), *patior, pati* ("undergo, experience"), and also several late Latin and medieval coinages designating mental activities, including *imaginari* and *fantasticari*. To indicate a state of becoming, deponency provided an elegant grammatical fudge to an otherwise strong logical binarism that too sharply divides passive from active states of imagining, knowing, feeling, and willing.

Topical Reasoning

In their standard logical definition at least from the time of Aristotle, Topics are "seats or places of arguments," *sedes* or *loci argumentorum*. The formulations are equally common, the word *sedes* being a literal translation of Greek *topoi*, and the word *locus* its equivalent synonym in Latin; *topica* is just the latinized form of *topoi*. But just what was meant by speaking in such terms? In what sense are "topics" seats? And just what is it that "sits" in such seats?

A very great deal has been written by historians of both logic and rhetoric on this subject.[9] The ancient works best known to most medieval writers were by Cicero, who around 55 BCE wrote on the Topics, in response partly to the eponymous work by Aristotle (*Peri topoi*) and partly to Stoic developments of Aristotle's ideas, but also from within his own profound knowledge and practice in Greco-Roman rhetorical teaching on the subject of Invention. Cicero's *Topica* in turn was commented on and developed further in the early sixth century by Boethius, and it is mainly Boethius's works that provided the immediate foundation for what medieval writers such as Hugh of St. Victor said on the subject, though Cicero's *Topica* itself was also known. And especially after the late eleventh century, Aristotle's work too was studied directly in Latin translations.

In *Topica*, Cicero defined these mental seats as follows: "As the discovery of matters that are hidden away is made easy by means of a known and marked location, just so if we wish to track down [=*pervestigare*] some argument we must know [the method of] 'places': for they were so named by Aristotle as 'seats,' as it were, from which arguments are drawn forth. Accordingly, we

9 Helpful discussions of *Topica*, with particular relevance to medieval thinking are Stump 1989, Marenbon 2003, Reinhardt 2003; discussions with reference particularly to literary invention include McKeon 1975 and D. Kelly 1978, 1998. Margolis 1987 studies the essential role of patterning in cognition, pattern-making being the basic tool of locational learning methods like those described in the various discussions of Topics and the method taught in so-called arts of memory.

may define a *locus* as 'the seat of an argument.'"[10] A Topic was understood more broadly then than we might now do, as indeed was what constituted an argument. Arguments included examples, maxims, stories, even omens and miracles, the complete contents of the florilegial collections of *dicta et facta memorabilia* that formed a core of ancient and medieval learning. We still search and find via topics online now, though we are encouraged to call them "keywords." Anything that might be persuasive to a particular group on a particular occasion was called an argument. Cicero, arguing for the defense, famously brought into court a disgraced Roman general, weeping over his battle wounds while the old soldier displayed his scars to the court – that bit of egregious theatre also was treated as an argument. Other Topics are still familiar to us: "who, what, where, when, why, how": these fundamental analytical questions were called *circumstantia* (they still are the basic analytical questions that "stand around" a convincing proof). Topics which served as cues and sources for Invention also could be particular words and phrases: justice, piety, love, know thyself, and so on.

Each such topic-seat can serve as the expandable "brief" for a sermon or meditation or other composition, and topics and sub-topics can be variously combined. But the main advantage of using such structures in composing and performing is that you will thereby know *where* your various arguments are, and can mentally go there directly and securely, without "losing your place," repeating yourself, and getting hopelessly "off the track." (Notice how embedded the idea of "place" and "investigation" still is in the way we speak of such matters.) You should also take pains that each seat should amply cue associative chains of other arguments that you have learned, so that you can call these forth as you choose, and then invoke, adapt, and elaborate them as you judge appropriate to your occasion. You will no longer be limited to performing your speech solely by heart, because relying on rote – as every rhetoric teacher since antiquity has cautioned – carries with it the distinct likelihood of forgetting *where* you are in your talk and being reduced either to a humiliating silence or to rambling about without any perceptible point or even discernible syntax.[11] John of Salisbury, citing Isidore, defines them well:

10 Cicero, *Topica*, 7–8 (LCL, trans. Hubbell):" Ut igitur earum rerum quae absconditae sunt demonstrato et notato loco facilis inventio est, sic, cum pervestigare argumentum aliquod volumus, locos nosse debemus; sic enim appellatae ab Aristotele sunt eae quasi sedes, e quibus argumenta promuntur. Itaque licet definire locum esse argumenti sedem."

11 The clearest medieval exposition of how to prepare such a structure of linked memory "seats" for one's own inventing is Hugh of St. Victor's preface to the *Chronicon* (time-line) of biblical history that he compiled for his students, translated as Appendix A in Carruthers 2008 and Carruthers and Ziolkowski 2000.

they are called "topics" because "they comprise *topo[i]*, that is 'places,' which are the seats for arguments, sources of our sense-based understanding, and authors of what we say. For this reason the teaching is also called 'Topics,' because it constructs (*facit*) the knowledge of 'places.'"[12] It is of interest, to me at least, that John uses the verb *facere* here and not the more usual *docere*, underscoring the active instrumentality of the Topics, not simply their received classroom content.

I have written extensively about the cognitive efficacy of modeling inventive thinking in this way. I stress that the practice (for it is a *practice*, a *techne* – you must perform and frequently exercise it, not just read about it in a book) embodies the recognition that memory and remembering are the foundation of all invention. The matters one seeks out in one's mind are thought of as "covered over," *abscondita* as Cicero wrote, an idea synonymous with another common Latin word for "forgotten," *oblitus*, literally "smeared over" and thus hidden from view. Another Latin synonym is *lapsa*, for forgotten things have "fallen" from our active minds and need to be uncovered, found and re-membered, their parts put together anew in our present thinking. To find an argument, one has first to investigate its tracks and traces, rather as one must do to find a rabbit hidden in its burrow. An image (Figure 11) from a fourteenth-century English psalter wittily recalls what one needs to do in order to meditate fruitfully upon the Psalms. It is but one of a great many such images in devotional books on the theme of locating and tracking down "prey" in order to nourish one's "praying" (the verbal-visual pun works especially well in English). Rabbits are also notably fertile creatures, so when they are dis-covered – the essential task of Invention – each one can generate much new food for our minds to chew over. Augustine remarks (in Book XI of *De trinitate*) that by such recollective searching through one's conceptualized experiences, one can intelligently comprehend not only things that one has never actually experienced but also things that cannot exist in Nature.[13] In this way, the Topics will indeed serve as sources and origins of how we (re-) cognize our world, and of the things we say about it. This model of human Invention underlies all medieval understanding of how human beings think and understand their ideas, even those about God. It is also crucial to grasp that any "location" is always determined and discovered *in relation to* other locations: its "place" or "seat" is not an essential quality of the thing itself.

12 John of Salisbury, *Metalogicon* III.9: "quae [topica] ut ait Isidorus sic dicta sunt quoniam topos continent, id est locos qui sunt argumentorum sedes, fontes sensuum, et origines dictionum. Pro eo autem disciplina ipsa dicta est topice, quod locorum doctrinam facit." (My translation)

13 See Augustine, *De trinitate* VIII.3–5, IX.2 and XI.4, and also Carruthers 1998.

This means that *content* relevant to an argument may very well be made discoverable through multiple associating chains and – unlike our habituated rote memorizing – is not found only in one form in one location, thence proceeding solely along one repeated path like a wheel (*rota*) in a deep rut.

One other matter concerning the medieval understanding of Topics is important to keep in mind. Roman and medieval teaching distinguished two technical verbal practices which had different goals, but both of which involved using Topics. They are best thought of as two functions of a basic analytical technique: as Boethius says in his Commentary on Cicero's *Topics*, "orderly reasoning has two tasks, judging and inventing."[14] The former, as Cicero says (citing Stoic ideas), is more formal, and involves analyzing the propositional relationships that are required when judging things to be true or false: this is what we now think of as "formal" syllogism. It is taught in Dialectic. The other task – inventing – also involves the method of Topics, but its goal is persuasion, and its arguments are more often *ad usu*, drawn from a particular society's common cultural wealth: its customs (*mores*), its shared knowledge (the *dicta et facta memorabilia* of a culture), and all its traditions and conventions, including of language and the arts.[15] In this context topical argument is taught as an aspect of Rhetoric also, even as its primary aspect. So there are two goals, *iudicandi* and *inveniendi*, that use the same art, the essentially rememorative methods and techniques of associating and locating *what* one learns, *where* one has put it, and *how* to find it promptly again. The method of Topics provides a map to the catalogued library of your mind, an effective way (when planned carefully and consulted frequently) for re-discovering even what you once learned but now has become obscured to your mental sight. And, as such, medieval *memoria* is the crucial bridge uniting the (material) experiences formed from our senses with the (immaterial) reasoning of our intellect. Or, as Isidore of Seville has it, *memoria mens est*, *memoria* in its broadest sense is Mind, the matrix of our thinking and knowing.

14 Boethius, *Commentarii in Ciceronis Topica*, Bk. I, praef: "omnis ratio diligens disserendi duas habeat partes, unam inveniendi, alteram iudicandi"; the original text is Cicero, *Topica* 6–7.

15 In Dialectic, this was called "enthymeme," its persuasive powers recognized as well in rhetoric. Though sometimes spoken of dismissively as incomplete syllogisms (which since their middle term is often unexpressed, they technically are), they are not necessarily only partial or wrong as ethical truths: see Burnyeat 1996. See also Leff 1983; Fredborg 1978; Stump 1989; on the shifts in early modern teaching, see Ong 1958; and on Valla and Agricola, P. Mack 1993. Rosier-Catach 1994 discusses medieval versions of modern speech act theory.

Imagining and Reasoning

In Passus 15 of the B-Text of *Piers Plowman*, Will encounters a strange creature, "a sotil [rarified] thyng," he says, "withouten tonge and teeth" which yet speaks.[16] Will isn't even sure it is "Cristes creature" (xv.15). Oh yes, it says, "I am Christ's creature," and it has many names "in Christ's court" according to its various abilities: *Anima* when it gives life to the body, *Animus* when it wills and chooses; *Mens* when it knows; *Memoria* when "I make mone [moan] to God" in confessing my sins; *Ratio* when it judges and does what truth teaches; and its first name is *Sensus* [as sense-derived knowledge] "and that is wit and wisdom, the source of all crafts." All these soul-terms were commonplace for centuries, a cluster of definitions straight from Isidore of Seville's encyclopedia (*Etymol.* XI.1). It seems a complete list of cognitive processes, yet there is one glaring exception – neither Isidore nor Langland includes imagining among Anima's functions, no phrase such as *dum imaginet, imaginatio sum* [when I imagine, I am imagination].

Which we now may find very odd. Not so much so in Isidore's time (early seventh century), since the late Latin deponent verb *imaginari* was then still found only rarely, having been first used around the time of Tertullian (d. ca. 220), and so not used by the ancient writers studied in monastic schools, nor in the Augustinian cognitive analysis that Isidore would have best known.[17] But still, it is odd for Anima to leave it out, since Imaginatyf is a character that Will has already encountered in the poem, three passus previously as another bewildering interrupter of his visions. In Passus 11, Ymaginatyf delivers the last in a series of rebukes that Will endures from Clergy, Dame Study, and other figures, including the emperor Trajan, who have each tried to teach him the doctrinal basics of faith. Will is a willful student indeed, and his pugnacious argumentativeness finally earns him not only a rebuke but banishment: he is summarily ejected from his own vision, and complains: "wo was me thanne/ that I in meteles [dream] ne myghte more have yknowen" (xi.404–5). But never mind, he says to himself, at least I now know what Dowel [doing-well] is. Oh really? says "one that looked on me" – well you certainly haven't

16 William Langland, *The Vision of Piers Plowman*, ed. A. V. C. Schmidt. My quotations are all from this edition, and translations are my own.

17 Boethius states activities of the soul as imagination, reasoning, and intelligence, conflating *imaginatio* and Augustinian penitential *memoria*, and linking *memoria* with desire and conscience much as *Anima* does here (and *Piers Plowman* does more generally). I discuss this further later in this chapter. *Imaginari* becomes frequently used only after the twelfth century, with the Aristotelian-Avicennan psychology taught in the universities.

done well in speaking so rudely to your teachers (xi.410–20). As the stranger starts to walk away, Will catches up to him, and asks his name – and it is "Ymaginatyf." He defines his function in Isidorean style: "ydel was I nevere,/ Though I sitte by myself, in sikness ne in helthe" (xii.1–2). He tells Will to stop arguing with Clergy and Study as though he were only in academic disputation with them, to stop "seeking after the whyes" of Nature (xii.216), and to amend his life.

How, then, does Langland's Ymaginatyf acquire its constant activity, its mobility (for it is never idle, it says); its rational virtue as Will's teacher; its judging ability, able to call Will away from asking all those pointless "whys" to which only Kynde [Nature] has the answer, and instead to think on his own life in order to amend it? How could such an Ymaginatyf even be an instrument for Will to amend his life? Will has already been shown his past life in the "mirrour of Myddlerthe" several hundred lines earlier: so why is he required to think on his past again here? Also, Reason has already appeared in the poem, arguing with Will before he encounters Ymaginatyf. Ymaginatyf's appearance *after* Reason is out of the scholastic Aristotelian cognitive sequence, if he is Aristotelian *vis fantasia*, the mental ability to fashion cognizable images from sense data in aid of rational recollection. Will certainly has been using *fantasia* in his previous visions (indeed making images and *visiones*, those being the products of the fantastical ability, is the mechanism of the whole poem). But this Ymaginatyf is not the characteristically Aristotelian, scholastic, university-trained *vis fantasia*, but another kind – a more Victorine and Cistercian *vis imaginativa*, founded on the speculative monastic practices of "inward-seeing" and contemplation described by Cassian and Augustine, and more recently (to Langland's time) and in a refreshed manner especially by Richard of St. Victor.[18] And, although Aristotle was definitely on Richard's mind when he wrote about contemplation, it was not in a wholly uncritical manner.

The Langland scholar Hugh White has linked Ymaginatyf to Richard of St. Victor's works on speculative thinking – this identification is, in my opinion, correct.[19] Richard notably distinguishes what he calls *officina Aristotelis* and

[18] On the exercises and method of monastic *lectio divina*, see Leclercq 1982, 1963. The English word "speculation" has greatly changed its meaning over the centuries, from sensory "seeing" both actually and mentally, to its common meaning now of "guessing" and "taking chances." The OED, s.v. *speculation* charts these changes from the fourteenth century on; perhaps unsurprisingly, they coincide with the rise of capitalism, stock markets, and the like.

[19] White 1986. Also see Leclercq 1982, 1963.

officina Salvatoris ("the Aristotle workshop" and "the Saviour's workshop"), a pairing which he does not present as wholly binary but as supercessive, the latter enlarging and surpassing the former. *Officina Aristotelis* pursues Knowledge, but *officina Salvatoris* raises human seeing into Wisdom. One should also conflate Langland's Ymaginatyf with Augustinian *memoria*, "making moan to God" in order to save his soul. And indeed that is rather a good characterization of the set of disciplines known among the early monks as *mneme theou* or (in Latin) *pia memoria* or just *memoria*. It includes the discipline of meditative reading known as *lectio divina*, and also, importantly, the exercise of examination of conscience, both necessary preliminaries to ecstatic prayer itself (if one might be so illuminated and graced by God).

These exercises are sets of disciplines, practices, and procedures. They are not single actions – this is important to understand. Nor should one conceptualize this model of the knowledge process mechanistically, as an assembly line of separate, if cooperating cognitive workers, each one fashioning and then attaching its particular bit into the final product we call "a thought." The model of human cognition primarily in use throughout the Middle Ages (and inherited from antiquity) is that of conception, gestation, and childbirth. The word "concept" itself embodies this model, deriving from Latin *concipere*, "to be pregnant." We still speak of our "embryonic" ideas and notions, to which we hope to "give birth." The ultra-rationalist, mechanical, factory model of thinking developed much later in European thought, along with the split of "reason" from "emotion." Concomitantly, there has been a telling change of emphasis over time in the lexical reference of the English word "faculty" (a noun born ultimately from the Latin verb *facere*, "to fashion"), from being the name for an *ability* to perform particular activities, to a name for the human *agents* who perform that activity, and now is applied commonly even to the buildings in which those agents do their work. That switch, gradual as it was, has brought about a crucial alteration in how we normally understand the word "faculty."[20]

The Latin Avicennan description of the operations of *anima sensitiva*, firmly based in Aristotle's account of it in *De anima* and its attendant treatises, uses particular terms to distinguish the various cognitive abilities that were considered to process the material data from the five external senses,

20 On the profound grammatical shift that occurred in these Latin terms during the later thirteenth–fifteenth centuries, from indicating various powers and abilities to act in particular ways (as *vires*, *virtutes* and *potentiae*) to indicating the agents who perform the actions, and from diagramming these as areas within a single, open, connecting chamber in the brain to visualizing them as particular actors housed in separate "cells" of the head, see Carruthers 2019.

all gathered up in the brain's *sensus communis*. These mental operations are *fantasia/formalis*, *imaginatio*, *estimatio* (sometimes also called *discretio* or *discernens*, "discernment"), and *memoria*.[21] These are phases of the single process that generates mental concepts from raw sense data. A useful (and now often reprinted) diagram was drawn in an English manuscript from the mid-fourteenth century (Figure 3). The diagram shows the perceiving and conceptualizing process as occurring in a single cavity in the brain, running horizontally (this accorded with contemporary medical thought) from *sensus communis* in the fore-part to *memoria* in the hinder part, its middle area dominated by *cogitatio*, "cogitation," the synthesizing and ordering action of receiving and shaping various external sensory data into a single multisensory *imago*, discerning and judging what it is that they show (this was called their *intentio*), and thus fashioning from them thought-objects (*fantasmata*) that are also made recollectable and refashionable, so that our reasoning intellect (*intelligentia*) could work with them *as we may wish and choose* at later times to construct new thoughts and reconstruct older ones.

The harnessing of imagining to the processes and goals of reasoning is accomplished through remembering. In Figure 3, note especially the dragon-like creature placed in the brain channel (called a *nervus*) between *vis cogitativa/imaginativa* and *vis memorativa*. Such corporeal *nervi* (the original of English "nerves") were thought to be hollow channels that carry the movements of the body's animating "spirits" (*spiritus*) that are the means of communication both within the brain and between it and the body more generally. These spirits, like all created things, have some mass (just as it says in Wisdom 11: 21), however rarified it may be, and thus they both move and can move other things.[22] Their motions are caused by some "affect," *affectus* – which can be from external or internal causes such as remembering past experiences. These "spiritual motions" pulse through the body's *nervi* like the ripples in a pond (to invoke a metaphor common in these texts). And such movements are constantly within our bodies, for our brain never stops moving, even when we are asleep. It is, as John Cassian said, like a great mill-wheel, forever grinding up whatever is fed into it, good and bad, lies

[21] I have discussed this psychology many times: see Carruthers 2008 and 2019 esp. The text that the drawing in Figure 3 illustrates is a Latin summary of Avicenna's *Naturalia* Bk. 6, also known as his *Liber de anima*.

[22] *Spiritus* has a range of meanings in medieval Latin, from simple "breath" to "life energy," equivalent to the Greek *pneuma*, which also has various medical and psychological functions: see Chenu 1957, Jacquart 2020. G. A. Smith 2008 offers a particularly vivid biophysical account of what constituted late antique *spiritus*, including its mass and measure.

and truths, for in itself it has no discrimination.[23] The spirits are also thought of as expressions of the soul within the entire body, not only of the mind and reasoning, for *anima* (soul) is the whole of the energy that gives life to our flesh, its fundamental principle.

Reasoning – any sort of ordered thinking – needs a tool to control the flow of our conceptualized thought-objects when they are summoned for any intellectual activity. This instrument of the mind was called the *vermis*, and in standard Galenic medicine it was considered to be a corporeal tool that, because our cranial movements alternately thicken and lengthen it, could be used to regulate the "spiritual" movements of our recollectable perceptions. In this context, the medieval Latin adjective *spiritualis* is nearly synonymous with our word "mental," a word whose Latin original, *mentalis* (the adjectival form of *mens*), comes into regular use only towards the end of the Middle Ages. Early medical texts observed that people commonly raise their heads when seeking to remember something, and lower them when reasoning: these acts were thought to engage the *vermis*, opening to let memories out for use when intellecting and closing to shut them off. The *vermis* is the gatekeeper of mental images during all conscious recollection. But during sleep or if the brain is addled, the *vermis* will not function rationally, and images can pour forth helter-skelter as hallucinations and dreams. In Figure 3, the dragon-like creature situated just above *memoria* is a visual pun for the *vermis* (medieval English *worm* also meant "dragon").

This diagram is in a miscellany of materials (Cambridge University Library MS Gg.1.1) composed chiefly in Anglo-French and Latin, though some are in English too. The most recent catalogue of the illuminated manuscripts in Cambridge libraries dates it to about 1330 (the beginning of Edward III's reign).[24] It contains some romances, prayers and other pastoral material, some land accounts, a Brut (a history of Britain) written in French and followed by a poem on the death of the King Edward I, also in French, providing the sort of eclectic materials one would expect in a large book, written at various times and used on diverse occasions in an English landed household at that time. And it also contains (just before some excerpts from moral commentaries on Job) a short digest of Avicenna's account of sensory cognition, accompanied by this small picture.

[23] I discussed the mill-wheel image at length in Carruthers 1998; since Cassian's work was specifically recommended by St. Benedict, the image of the mill-wheel became traditional in monastic and later thought. A contemporary use of it, translated into modern cognitive psychology, is in Heyes 2018.

[24] Described by Binski et al. 2011, item 149 (pp. 139–41).

It diagrams only a standard Avicennan account, summarizing and commenting on Aristotle's account of memory, of how a mental representation (*fantasm*) of sensory perception comes to be conceived and stored in a way that makes it able to be later recalled. But notice that the drawing does not include any remembering process – the terms for recollection (*reminiscentia, recollectio, vis rememorativa*) appear nowhere in it. Nor do forms of *ratio* or *intellectus*. Instead, forms of the Latin adjective *imaginativa* are recorded as alternative terms for each phase in the perceptualizing process, together with its general name *imaginatio*, all deriving from the deponent verb, *imaginari*. In other words, this is a diagram demonstrating a pre-modern understanding of what we now call, with a misleading definite article, the Imagination. But this picture shows imagining (*imaginativa*) as a complex of activities that enable us to conceptualize the data our senses receive from outside and to recall it in a way that our minds can use. The picture shows schematically how such data from the various external senses are received into the brain (in "*communis sensus vel fantasia*"), and are combined into something that we cognize as a coherent shape (*forma*). This latter action is labelled "*ymaginatio vel formalis*." We also make an initial definition (or "estimation") of our perception, judging its source as hurtful or beneficial, to be avoided, ignored, or approached. Some of these perceptions then consolidate in our brains – after some short period of time – as what we call "memories," notions that remain in some way accessibly re-collectable when we are actively reasoning. These wholly mentalized notions are the embryonic materials from which, having recollected them, we can "conceive," shape, and thus "develop" our ideas. This is a wholly gestational and generative model, not a mechanical one.[25]

In Scholastic discourse such an experientially based concept was sometimes called an *intentio*, sometimes a *notitio*; I tend to use the English word "notion," which also works well in French, *notion*. Aristotle called these mental constructs *noeta*. Philosophers of mind sometimes use the term "thought-objects." Whatever word is used, a notion is not yet a memory (and has a way to go before becoming an idea, let alone a theory). Only after the cognizing and imagining process is completed (in the diagram, during *cogitatio vel imaginativa*), and then is finally processed by *vis memorativa* does a notion become available for active re-collecting (*memini*).[26] For this

25 I describe this process in terms of this diagram particularly in Carruthers 2020 and 2019.

26 The way this stage is modeled is remarkably like what in contemporary neuropsychology is called "consolidation," which recognizes that memorability is an attribute that is acquired after some (usually brief) time has elapsed between our

reason, all remembering is of necessity *praeterita*, of "events past" that are no longer immediately present in our external senses. A medieval dictionary, the *Derivationes* of Uguccione of Pisa (from ca. 1190), repeats Isidore of Seville's derivation of Latin *mens* and *memoria* from the verb *memini*, and then goes on to say that *memoria* is also related to Latin *mora*, a "delay":

> Memory is the stable lingering in our mind of our cognitions, whence indeed the word is derived from "delay"; for it is also said by etymology that *memoria* is a kind of delay within the mind.[27]

In this generative psychology, which is basic to the medieval philosophy of Soul and Body (*de anima*), recollection is always a re-making, re-membering, re-collection. A recollected notion always had to be re-imagined and "collected together again" when it was reconstructed; it was never the same as the mind initially produced. This is fundamental to comprehend. One important consequence of this idea is that mnemonic "accuracy" and "exactness" were a different issue – differently based, and differently evaluated – for medieval scholars than for us. In medieval accounts, a predisposition to invent is built into the human recollecting process itself. Remembering can only be an action of trying to re-make in the present, specifically by means of a search starting from what you know now, and reasoning by a chain of remembered associations until you come upon (*in-venire*) what you are looking for.

As a method of thinking, this was commonly known as ratiocination (*ratiocinatio*), a term familiar from Roman logic and rhetoric throughout the Middle Ages. Cicero defines ratiocination generally as reasoning from common places and topics and thence "to proceed by ratiocination from that which is written [for example, in a particular law text] to what is not written [extending the law to analogous new cases]."[28] He also defines *ratiocinatio*

---

immediately received perceptions and our ability to recall them. See Dudai 2002; Roediger, Dudai and Fitzpatrick 2009, s.v. "Consolidation," also "Retrieval" and "Working Memory."

27 Uguccione da Pisa, *Derivationes*, s.v. *Memini* (M77), #7: "[E]st memoria firma comprehensorum mora in mente, unde et quidam derivetur a mora; quod autem dicitur memoria, quasi in mente mora, ethimologia est." (My translation, from Cechetti ed. II. 749).

28 *De inventione* II, 50.152: "a ratiocinatione oportere coniectura ex eo quod scriptum sit ad id quod non sit scriptum pervenire"; LCL, trans. Hubbell (I cite Hubbell's translation throughout my discussion). Hubbell consistently translates legal *ratiocinatio* as "reasoning by analogy" with similar cases. But *ratiocinatio* was not confined solely to formal reasoning; Hubbell notes (p. 92) that Cicero applies the term quite loosely, to what Aristotle identified as enthymeme, not just to strictly formal syllogism.

as a method seeking truthful demonstration not through resemblance and difference (*mimesis*) but from within the matter itself, "a form of argument which draws a probable conclusion from the fact under consideration itself [and which] is set forth and recognized in itself and may be proved from its own import and reasoning."[29] Ratiocination is a logical method of mapping out an itinerary, a journey in stages from what you know now (that is, starting from some knowledge you are aware of right now) to discover via a chain of associations what you have forgotten. Notice how often the verb *venire* is used for ways and modes of thinking – *pervenire*, *invenire*, *revenire*, *advenire*. *Via rationalis*, a "reasoning way," is a common medieval idiom for *ratiocinatio*.

The role in reasoning of imagining and remembering was also discussed well before the thirteenth-century Aristotelians in the works on contemplation by the Victorines and other twelfth-century scholars, including Thierry of Chartres (d. 1150) and the Cistercian Isaac of Stella (d. 1169).[30] Richard of St. Victor's introductory treatise on preparing for contemplation, *De duodecim patriarchis* (also called *Benjamin minor*), discusses the roles of the various human mental activities in terms of an allegory of the wives, concubines (*ancillae*, "handmaids"), and children of the Hebrew patriarch Jacob, he who was granted a vision of a great ladder stretching to heaven, upon which angels descended and ascended, and from which God spoke to bless Jacob's fruitfulness and future lineage (Gen. 28: 10–22).[31]

The Twelve Patriarchs is interesting as a work *de anima*, and not only because of its detailed account of Victorine spiritual discipline, important as that is for all later mystical thought.[32] But of particular interest to me in his discussion is Richard's account of the relationship of Imagination and Reason,

29 *De inventione* I, 34.57: "Ratiocinatio est ex ipsa re probabile aliquid eliciens, quod expositum et per se cognitum sua se vi et ratione confirmet." *De inventione* (called the *rhetorica antiqua*) was the elementary rhetoric text in schools throughout the Middle Ages; these definitions of *ratiocinatio* are also emphasized in the late antique commentaries and manuals by Marius Victorinus and Julius Victor.

30 Jeauneau 2020 offers a valuable reassessment of the School of Chartres, and the fruitful relations among these twelfth-century scholars mainly in Chartres and Paris, but also in the cathedral school in Liège, about which see also Renardy 1979, Riché 1989.

31 I discussed this scene, and the meanings of *terribilis* and *horribilis* in similar biblical scenes of divine vision in Carruthers 2014.

32 Richard's treatise has been translated by Zinn (1979); a new edition of the Latin text by J. Chatillon et M. Duchet-Suchaux was published in 1997; this supercedes the text in the PL. All further quotations are from this edition, with my translations after profitably consulting Zinn's.

or "Bilah" and "Rachel" as Richard denominates them. Rachel (Reason) is one wife of Jacob (who is both the Visionary and Patriarch), his other wife is Leah (Affection, Desire); both are needed for true contemplation. Each is attended (as they are in Genesis) by a handmaiden (*pedissequa*, *ancilla*), with whom each must work but also must restrain and guide. Leah's handmaid is Zilpa (Feeling), and she is all bodily delight; without the guidance of Leah she is swollen with pleasure and drunk with wine all the time. Rachel's handmaiden is Bilah (Imagination), whom Richard characterizes as *garrula*, a voluble chatterbox. By herself, Rachel is barren (as was the biblical matriarch); she requires the womb of Bilah in order to be fruitful in her reproductive role within the family (note again the conception, gestation, and birthing model that lies at the basis of this psychological allegory): "[f]or without imagination, reason would know nothing" just as "without feelings (Zilpa), affection (Leah) would understand nothing."[33] This allegory is compatible with later Scholastic psychology, namely that the human mind requires sensation-derived mental *imagines* in order to know anything, that imagination works with reason, and learning with remembering, and that knowledge also requires both a desire and *willingness* to know and recall a particular matter, for we know what we wish and choose to know. But the Victorines derive these ideas primarily from Augustine and Boethius. They are qualities also basic to the common ideas in antiquity of what is required to bring about rhetorical persuasion and *fides*, confident belief.

Imagination, Richard says, is so noisy within our minds, constantly clamoring visually and audibly, that Rachel is barely able to live with her. Even when Rachel retires to her solitary chamber (*cubiculum*) to read and study, meditate, and contemplate, Imagining goes on and on incessantly. When we try to pray and recite the psalms, our voluable "Bilah" recalls to our inner eyes every image and fantasy we have ever seen or heard, which she recollects over and over again, and never ceases repeating herself, though we ourselves may desire not to listen to her. She is like those old men and women who tell the same stories over and over, even though no one is attending to their chatter.[34] Yet, since Rachel, as reasoning, requires Bilah's service to be able

33 *De duodecim patriarchis* ch. 5: "Nam sine imaginatione, ratio nihil sciret; sine sensualitate, affectio nil saperet." Chapters 3–6 contain Richard's basic discussion of these four powers and their relationships: imagination and reason, sensory experience and affection; see also chs. 17–24.

34 *De duodecim patriarchis* ch. 6: "Est enim Bala [Bilah], garrula; Zelpha [Zilpa], tumulenta. Bala namque loquacitatem nec ipsa ejus domina Rachel compescere potest.... Imaginatio autem cum tanta importunitate in auribus cordis perstrepit, quatenus ejus clamorem, ut diximus, ipsa Rachel vix vel omnino cohibere non

to think at all, Bilah's unruly behavior is a destructive distraction for her, and requires not only strict direction and governance by rational method but also (and often) divine help in the form of truthful "illumination."

Imagination and the Passions: John of Salisbury's *Metalogicon*

It is concerning the role of mental illumination that Richard's analysis, and twelfth-century Augustine-based Victorine/Cistercian psychology more generally, differs the most from the Aristotelian-rationalist psychology of the academic schools. Rachel is not just human reasoning but, as the first wife of Jacob the visionary, she is "ratio divina revelatione illuminata [reason illumined by divine showing]."[35] A somewhat similar analysis of Imagination is in John of Salisbury's *Metalogicon* (ca. 1159), composed around the same time as Richard of St. Victor's *Twelve patriarchs*. The *Metalogicon* is a treatise on the language arts of the trivium and the verbal skills and ethical and civic virtues that they foster and sustain, more than on the soul and its powers as the matrix and necessary tools of contemplation, and for John of Salisbury Imagination and Reason have a necessarily synergistic relationship to one another. "Imagination," John says, "dawns [*oritur*] from the root of our sensations by means of the kindling-wood [*fomes*] of memory, and by a particular shaping [of them] together [*simplasim*], which we can term a conformation [*con + forma*], it sees inwardly [*intuetur*] things not only present but also absent whether in place or in time."[36] All learning, all experiencing, is

---

possit. Hinc est quod saepe dum psallimus vel oramus, phantasias cogitationum vel quaslibet imagines rerum ab oculis cordis amovere volumus, nec valemus.... Quaeque etenim visa, vel audita, quandoque quae ipsi gessimus vel diximus, ad memoriam revocat, et quae ipsa jam enarrando explicaverit, eadem iterum atque iterum replicare non cessat." My English paraphrase of the Latin, after consulting Zinn's translation.

35 *De duodecim patriarchis* ch. 4 (trans. Zinn, 57).

36 John of Salisbury, *Metalogicon* IV. 10: "Imaginatio itaque a radice sensuum per memoriae fomitem oritur, et non modo praesentia sed et absentia loco quidem uel tempore per quandam simplasim, quam nos conformationem possumus dicere, intuetur"; my translation. Note John's use of the deponent verb, *intueor* (modern English "intuit"). The unusual Greekism *simplasis*, appearing here in the accusative singular form (difficult Greekisms being a favored twelfth century learned trope), compounded from Greek *syn* and *plasis*, "mold, shape," emphasizes the action of fashioning materials, like a carver or metalworker; see DMLBS, s.v. *symplasis* (this occurence is the only one this dictionary gives). Since the *fomes*, "fuel," most commonly to hand was wood-kindling, it is also tempting to see in John's very learned metaphor an allusion to the *silva*, the "wood" that is "shaped" artfully

analyzed as various sorts of contingently related sensory movements in time, from "there" to "here" and equally from "then" to "now."

These imaginings are made from stimuli received in our external senses that set up physical movements carried to our receiving *anima*, "soul" (though here perhaps "life-force" might be a more precise equivalent). John describes them as pulses: "In the first place, imagination is a movement of the soul pushed [*pulsatae*] from the outside."[37] John's world, in keeping with the physics of his time, is one of continual motion and change, and thus pulsing motions carried through the air or other elements are the means of all our sensations. In the form of such movements, sense data then move through channels filled with the spirits (*spiritus*) that animate the body generally and into the *communis sensus* of the mind, where imagining fashions them into the *phantasms* with which thought works.

The soul's movements are also called *passiones*, "passions," and *affectiones*, "affects," a word derived from the Latin root *facio*, "make, shape." *Affectiones* has a double meaning, simultaneously the effect on our sense-organs when they are struck by various pulses from the world outside, and also as the "feeling" that such "impulsive" movements arouse in us. But these sensorily derived imaginings, whether immediate or created later from remembered fragments, are each themselves a mixture of often contrary pulses in us, tugging in opposite directions – some hopeful, some fearful, some positive, others negative and even wickedly destructive. As hope flourishes, sadness and grief do too, and if these feelings grow in strength, they disorder and roil the soul, and anger comes from their clash. There is no way that imagination acting alone can sort, order, and turn them to virtue.[38] John gives two examples from classical literature of how imagination can concentrate distressed feelings in a soul, leading even to violence. The first is the story told in Ovid's *Fasti* ii, of how his constant imagining of the chaste Lucretia drives Tarquinius to rape her at sword-point, and the second (in *Aeneid* iii) is Aeneas's encounter with the exiled, distraught Andromache who exclaims, in tears, how the Trojan

from one's various experiences re-viewed by means of our memory (just like a dawning sun, *oritur* being an unusual verb to use in this context). I have used the CCCM edition of the Latin text (1991); the translation is mine, closely consulting that of McGarry 1955.

37 *Metalogicon* IV. 11: "imaginatio primus motus animae extrinsecus pulsatae."

38 *Metalogicon* IV. 10: "Mixta autem est ex contrariis, cum spes fruendi uoluptatem pariat, sed dilatio tristitiam, et dolorem. Quae si inualuerint, ut exordinent et turbent animam, ex ea concussione fit ira." On Aristotle's analysis of mixed feelings in rhetoric, see Frede 1996. In Carruthers 2013, I analyzed this Stoic-influenced model of emotions in terms of Chaucer's characters in *Troilus and Criseyde*.

wanderers she now sees before her remind her of her murdered son Astyanax, whose image, John says, she has drawn forth mentally in her recollection ("*imaginem abstraxisse recordatione… fatetur*").[39] John's account of someone abstracting an image recollectively from a past experience into the present as a mental image is nothing remarkable. But what is worth noting is that both of John's examples are of imagining performed in states of great distress and disquiet, and when the person is also removed from their ordinary, habitual place, Andromache from Troy, Tarquin from the battlefield. Both accounts stress these anxious, disorienting conditions, ones that are also conventional, as we have seen, for any creative thought (whether their thoughts, when fully formed, are turned to virtue or to destruction).

John's emphasis is on how disturbing to a person's whole animate being, not just mentally but corporeally as well, these *passiones* invading from outside can be. It is an image of primal chaos itself. A compatibly Stoic-Aristotelian psychology can be detected in his account, but it is much more vehement than what one finds a century later in academic writers on this subject, such as Albertus Magnus and Thomas Aquinas. An analogy to the turbulent physics of a Neoplatonist, almost Calcidian, account of creation seems to me to drive his energetic description of how Imagination works; by comparison Aquinas seems always in control of all his matters. And of course it is precisely the unwavering logical form, the *ordinatio* of the *summa* genre which Aquinas perfected, that creates the impression of calm certainty on which the persuasiveness of his work relies.

The virtue which can turn, teach, and order our chaotic imaginings positively is Prudence, *prudentia*, which had a much more inclusive meaning for John of Salisbury than it often has now, where it is used reductively as a synonym of thrift. I have written extensively about the reliance of Prudence on Memory – indeed, through the seventeenth century, *prudentia* was considered to be an aspect of *memoria*, the face that looks to and plans for the future.[40] Sensory knowledge is very fallible, and cannot be at all depended on to know actuality from illusion. But Imagination has a second kind of movement, one which *wants* to know whether the sense images we have fashioned are of actual things or not, and judges how we should immediately respond to them. This mental capacity is like the scholastic Avicennan *vis estimativa*; we might

39 *Metalogicon* IV. 10. Note that *fatetur* (from *fateor*) is another deponent verb, both receiving (passive) and acting.

40 Carruthers 2008, pp. 76–89, 238–9 esp. The basic Latin discussion is Cicero, *De inventione* Bk. II, sections 53–54. John of Salisbury discusses Prudence's three temporal faces in *Metalogicon* IV.12.

now think of it as a basic human cognitive need for *shape*, for pattern and order, the syntax of our experiences, so that we can comprehend the world around us confidently. John identifies this as "opinion" (*opinio*), a category of Aristotelian logic.[41] It is a kind of reasoning but its conclusions are not universal and necessary: "And because (our mind) detects the falseness of our sense-perceptions, it moves vigorously within itself so that it also might grasp something believable, on which it might confidently rely without error. From this agitated movement is born the virtue which Greek speakers call *phronesis*, Latin speakers *prudentia*."[42] Prudent, wise judgment is first a matter of *desiring* to believe what is true and not false, involving consciously deliberated choices from among from the welter of our brain's incoming "affections," those feelings moved and stirred up by the *passiones* that, either immediately or *in absentia*, have "pulsed" through our animated spirits from our various senses.[43] John's is a remarkably material, corporeal account of cognitive perception (though less remarkable in his twelfth-century context). Notice also that the virtue known as Prudence is *born* within the mental agitation caused by the confusing swirl of sensations that have disparately affected the soul (*ab hac agitatione nascitur virtus*) – order is born from chaos once again. Prudence is also, as it is in Cicero, a virtuous ethical habit to be cultivated especially by orators (along with Justice, Moderation, and Fortitude), for confidence and trust (as *fides* and *fiducia*) give birth to ethical and civic Wisdom (the English equivalent of Ciceronian *prudentia*). John makes the Aristotelian distinction between contingent "social truths," the enthememicly argued commonplaces (*topica*) defining a particular society's behavioral

41 The concept of *estimativa* was greatly expanded and rationalized in thirteenth-century Avicennan Aristotelianism. John discusses it in Stoic terms as a matter of movements that are "first" (the sensory data received in pulses from outside), and "second," the initial actions of assessing and defining made by the judging *mens*. This assessing action is *fronesis/prudentia*.

42 *Metalogicon* IV.11: "Et quia sensuum fallaciam deprehendit, in eo agitatur ut et fidele aliquid teneat, cui sine errore fiducialiter possit inniti. Ab hac agitatione nascitur uirtus, quam Graeci fronesin, Latini prudentiam uocant." *F* is the common Latin spelling of Greek *φ (phi)* at this time. On Aristotle's account of experiential truth, *phantasia* and memory, see Frede 1992.

43 *Metalogicon* IV.11: "Est igitur imaginatio primus motus animae extrinsecus pulsatae, quo secundum exercetur iudicium, aut per recordationem redit primum." The "first" assessment is sensory, either immediate (such as sweet, sour) or by recalling a similar sensory sensation. The "second" is of the image only and is made through comparison with recollected images. These together form "opinion." The received sensory "pulses," *passiones*, are integral to a complex judgment in which "feelings" play a significant role along with reasoning.

norms and expectations at a particular historical moment,[44] and those truths which are always and universally necessary. *Prudentia* is particularly a virtue supporting social truths, *domesticae necessitates*, the truths of daily social, commercial, and political life.[45]

Like Richard of St. Victor, John gives Reason (*ratio*) a restraining role upon the disorder of our sensory perceptions, their *inordinati motus* as he calls them, which Reason, presiding in the citadel of the head, brings together, orders, and directs toward the good of all (*ad normam bonitatis componit universa*).[46] Like the Senate debating and legislating in the Roman Capitol, Reason sits between the brain's cells of fantasy and memory, and from her watchtower (*e specula*) she can assess the judgments of the senses and of imagination. But even Reason is not entirely dispassionate in her role, for at times she can get excited, fanned into flame by the bellows (*ventilabro*) of the mind's sensory experiences and imaginings. So Prudence also has her servant Philology to advise her judgments, and Philology in turn has two helpers, Carefulness (*Periergia*) and Watchfulness (*Agrimnia*). Carefulness is attentive to the *whole* of any work (*laborem circuit operis* – not cherry-picking her authorities or skipping about in the text) and Watchfulness's vigilance ensures that the effects of Reason's judgments are appropriate to the particular cases (and texts) to which they may be applied: her character accords with Aristotle's principle of judicial equity. Philology is Earthly and temporal in her linguistic origins, but with her helpers "transits [*transit*]" towards wisdom and knowledge of the divine. Since Prudence also concerns Earthly matters and the love of reason, together with all her helpers humans can surge in knowledge toward unchanging divine truth, in this way overcoming their mortal condition, however momentarily.[47]

44 John is careful to discuss enthymemic truths as conditioned historically as well as socially – he also is aware of the value of such truths in commercial dealings (*commercia*), a telling symptom of increased urbanizing and of foreign trade during the twelfth century; see *Metalogicon* IV.13. On the truth-value of enthymemic reasoning, see Burnyeat 1996 and 1987.

45 *Metalogicon* IV.12. These three areas, adopted from Aristotle's political and ethical works, constitute *philosophia moralis*, "moral philosophy," in later Scholastic university curricula. Public oratory, especially as preaching and diplomacy, came to be subsumed into moral philosophy as well, on the model of Aristotle's *Rhetorica*, developments well discussed by Copeland 2014 and 2007.

46 *Metalogicon* IV.17. All Latin quotations and my paraphrases in the following paragraph are from this section.

47 *Metalogicon* IV.17: "Habet autem terrenam et mortalem Philologia originem, sed cum ad diuina transit immortalitate quadam deificatur, quia cum prudentia quae de terrenis est et rationis amor, ad incorruptae ueritatis diuinorum que arcana consurgit, in sapientiam transiens, quodam modo a mortalium condicione eximitur."

Similarly according to Richard of St. Victor, one understands what is humanly *incredible* (the Incarnation, for instance) by means of what is *credible*, moving *per rationem* (through reasoning) to *supra rationem* (heightened reasoning) to *praeter rationem* (beyond reasoning). The term common to all these stages is *ratio*. This is not faith seeking understanding so much as it is prudential reasoning spread outward and upward in steps (*passus* in Langland, *gradus* in many other writers) towards confident faith via rational methods using the tools of recollective imagining.[48]

Vision, Illumination and Fantasy, Dante's "alta fantasia"

As he rises towards his final envisionings of Paradise and the Trinity at its heart, Dante describes how his mind is changing. Though he already has become "transhuman" (*trasumanar*), the rarified state in which he is able to rise up with Beatrice through the moving cosmic circles of the moon, sun, and planets, to the heaven of the fixed stars, he feels his mind elevating yet further. As the Light grows even brighter, as his mental sight becomes increasingly enflamed (recall the fiery nature of John of Salisbury's characterization of thinking), one point ("un punto solo," *Par.* 33.94) fixes and enkindles his enraptured gaze.[49] He looks more attentively, his mind transhumanly all motionless, fixed on that one point and then, as his mind alters again with his new visionary power, the illuminating Point alters also, resolving into the interlocking shapes of "tre giri," three wheels of different colors ever circling within and about each other and yet ever unchanging (*Par.* 33.116–17). In terms of this image his mind (imagination, memory, and reason working together) can now better understand the Trinity.[50] One wheel reflects the

48 In her study of desire in *Piers Plowman*, Zeeman 2006 has very well observed that "Langland so often thinks not in … a definitional manner but by analogical side-steps." Her discussion of Kynde and Ymaginatyf also emphasizes the Augustinian generative model of thought based in desire, esp. 245–57. See also Lombardi 2007 for a stimulating discussion of the relationships of desire, syntax, and grammar, especially in Dante.

49 Ed. G. Petrocchi, 2nd ed. 1994; all quotations from the *Commedia* are from this edition. The rarifying of Dante's thoughts into air and fire as its objects become ever more sublime is reminiscent of Clarembald of Arras's analysis of the process, discussed below. Dante's language here should not be understood as only metaphor or allegory, for there is physics in it also.

50 A fine interdisciplinary effort, complete with several possible mathematical models, to comprehend the *tre giri* as a ring topology is Saiber and Mbrika 2013, though they admit that their models do not fully account for the poem's description. On the physics of Dante's *Paradiso*, see esp. Boitani 2007.

second like a double rainbow, the third breathes ("si spiri") like fire equally from the other two. As he looks ever more closely, his mental gaze ever sharper, the pilgrim can make out a human shape reflected in the second wheel. He wants to see even more, his whole intellect concentrated to understand just how that human form is located within the wheels. But at this Point his mind fails, language and memory failing together – and the poem ends. Or rather, as Dante writes, "A l'alta fantasia qui mancò possa [Here high fantasy lacks its power]" (*Par.* 33.142). Without the imagining ability that his mind requires for understanding, only his willing, loving desire can move him now, with that same gravity-like pull toward and into Eternity that also moves the cosmos.

Certainly much has been written about this whole passage, and I do not intend to challenge any of the currently received interpretations of it. But it seems to me that the phrase *l'alta fantasia*, understood correctly as the pilgrim's ability to fashion images within his mind with which to try to comprehend divine Being itself, refers to a particularly Cistercian model of human intellect and the role of *vis fantastica/fantasia* in the process of shaping our embryonic concepts. This Cistercian model is not radically different in kind from the more familiar (to us) Avicennan–Thomist model.[51] But it gives a different emphasis, more Augustinian–Boethian (and Victorine) in nature, to the divinely granted illumination and enlightenment needed for humans to be able to understand the highest Truths.

The concept of *alta fantasia*, the power of mental envisioning illuminated directly by divine contact (an offspring of Augustine's concept of "inner light") is diagrammed in Figure 12. This drawing shows the process of conceptualizing as it is described in a mid-twelfth-century treatise *De spiritu et*

[51] Ongoing tensions during the thirteenth century between Dominican-Aristotelian-Avicennan and Cistercian-Franciscan-Augustinian writers on the soul are well known to historians, and were likely exaggerated (in the manner of so many academic quarrels, medieval and modern). Albertus Magnus dismissed *De spiritu et anima* as saying many false things, and Aquinas said its ideas were not truly Augustinian, but the unauthoritative, cherry-picked compilation of "cujusdam Cisterciensis [some Cistercian]." Other academics, including Bonaventure, who were kinder to its evident Platonism, received the treatise warmly. On its reception and attribution (often to Bernard of Clairvaux), see especially Mews 2018, who quotes Aquinas's dismissive words in p. 320 n117. Relevant also is Grosfillier 2013, who describes a similarly hostile reception from university Aristotelians including Aquinas, for parts of Richard of St. Victor's *De contemplatione*, discussed later in this chapter. McGinn 1977, p. 26, has characterized these divergent emphases as "naturalistic" (the Aristotelian-Scholastic) and "spiritualized" (the Cistercian-Victorine); see his summary discussion of the theological complexities involved, pp. 2–27.

anima, likely by Alcher of Clairvaux. The treatise in turn draws heavily on a somewhat earlier Cistercian treatise by Isaac of Stella on the same subject, as well as on the work of Thierry of Chartres.[52] The drawing accompanies an early copy of Alcher's treatise, though the picture itself was evidently made well after the text it accompanies was written (the scribal hand appears to be later thirteenth century, but the drawing on folio 47 likely dates to around 1400, perhaps somewhat later).[53] It was drawn crudely in red ink on the recto of a blank leaf in the original booklet, facing the final verso side of the written copy, which contains a university-style outline of the main points of Alcher's treatise. Outline and drawing thus make a single opening of the book, folios 46v–47. Though likely copied initially, possibly for personal use, in a Cistercian monastery located in northeastern France, by the later thirteenth century it had become part of the library of the recently founded (1248) Cistercian Collège St-Bernard in Paris.[54]

Shown on the recto of folio 47 of this manuscript are the five stages of the Avicennan conceptualizing process, with *sensus communis* in the anterior chamber of the head just behind the face (the organs of the external senses are all shown: *gustus*, *auditus*, *olfactus*, and *visus*, with *tactus*, shown as a hand at shoulder-level, near the heart). Next to *Sensus communis* are *Ymaginativa* and *Estimativa*, and in the rear portion is *Memoria sensualitatis uel Retentiva*, the retentive quality of our cognized, sense-derived notions of the world around

[52] Translated in McGinn 1977. Nearly every aspect of this treatise's attribution and dating has been disputed over several centuries, though it does appear to have been in circulation by the 1170s. Mews 2018 has carefully sorted the evidence, and concludes convincingly that the author is Alcher.

[53] Many thanks to M. T. Webber, P. M. Jones, and N. Bell for examining, at my request, the collation and scribal hands of the manuscript itself, now in the Wren Library of Trinity College, Cambridge, and for meeting with me there, the manuscript open before us, to demonstrate their findings. The first booklet, made up of six octavo quires (48 leaves), contains Alcher's text, which concludes on folio 46; on the verso side of fol. 46 is a scholastic-style outline of its topical *divisiones*, in a hand that Webber believes to be that of the same scribe as wrote the text in the mid- to later thirteenth century, not long before (or shortly after?) the book came to the Cistercian library in Paris. The writing on the facing page diagram (fol. 47, as in Figure 12) is later, and the rest of the material in the book is also fifteenth-century Cistercian in origin.

[54] The manuscript is quite small and written in a single column, not the double-columned large-page format more common in books made for library use. See Mews 2018 for a full description of this text as it appears in this particular manuscript, which Mews thinks is of particular importance in establishing it editorially. It was usual for monasteries to gift some of their books to larger, often newer, university foundations of the same order.

us.[55] But in a chamber at the very top of the head is *Fantasia*, located right under the middle of the skull, which is labelled *Interciput*, an unusual word to use in this context, as it is also very unusual to have a head diagram that separates fantasy from imagination (compare the Avicennan description shown in Figure 3). This crude drawing was certainly not made by a trained artist-scribe, as is plain from its exaggerated features and the way that the head (in profile view) is positioned on the shoulders (in three-quarters view). But it does reflect the Cistercian–Victorine Neoplatonist account given in *De spiritu et anima*, and in Isaac of Stella and William of St. Thierry, which derived anciently from Augustine, but more recently from Hugh of St. Victor and Bernard of Clairvaux, with whose spiritual works on meditation *De spiritu et anima* was also often associated.

The drawing shows *fantasia* as indeed *alta*, elevated as Augustine described, the mental path running from the face at the front, under the skull (labeled *Sinciput*), moving through the *imaginativa* and *estimativa* powers to "sensory and retaining memory [*memoria sensualitatis uel retentiva*]" in the rear area (labelled *Occiput*). *Fantasia*, a separate power, is positioned *alta*, "above" and on top of all the sensation-derived powers, in the uppermost skull area just under the scalp, the area closest to immaterial *intellectus* and *intelligentia* and so nearest to the directly illuminating (though invisible) rays, Augustinian *lux etherea*, with which ordinary human *visiones* can be divinely irradiated. The scalp also is the location of baptism, the sacrament that first marks and illuminates the soul with Christian grace.[56]

Augustine's own description of the soul's cognizing procedure is influentially given in Book VII of his treatise on the literal text of Genesis (*De Genesi ad litteram*), where he discusses the creation of Adam. There he argues that the animate soul (*anima*) administers (*administrat*) all the body's various movements, including all its sensory reception from the corporeal organs of sensation. The senses are the messengers of our cognition, received as various

55 From around 1300, Duns Scotus and later writers made a distinction between *memoria retentiva* (by which they meant recollectable, brain-based and sense-derived *notiones*) and *memoria intellectualis*, remembering the ideas made and understood by human intellects, abstracted mathematical and theological ideas among them.

56 *Interciput* (*inter* + *caput*) is a rarely occurring medieval word, though coined on the model of more common *occiput* and *sinciput*. A word search in the LLT records it once in the *Graecismus* of the grammarian Evrard of Bèthune (writing at the turn of the thirteenth century in north-western France, likely aided by several contributors) and once in the writings of Raymond Llull a century later. The top of the head was also called *cerebrum*, and associated in the liturgy with Baptism, as being particularly receptive to divine grace: see DMLBS s.vv. *interciput* and *cerebrum* d.

wave-like animate motions transmitted through our nerve-channels into the brain and there transmuted by corresponding inner senses into the images and fantasms that are the materials of conception and cognition. This, he says, is the opinion of doctors, *medici*. These doctors have described this mental pathway as proceeding from the face, where all the sensing organs save touch are located, into three *ventriculi* in the head. Latin *venter* can mean both "stomach" and "womb"; *ventriculi* are small versions of these, so the cognition process is modeled as gestation and also digestion. We not only "give birth to" our ideas but we also "chew on" our experiences before fully articulating them.

The anterior ventricle/womb receives all the sensory data from the various external organs including touch, through the body's *nervi* into the anterior brain. This is why, Augustine says, God breathed "*flatus vitae* [the breath of life]" directly into Adam's face so that Adam became a living soul (Gen. 2: 7). All sensory data is gathered and mentalized here into the motile thinking-images Augustine elsewhere calls *intentiones*.[57] The mental path terminates where the skull joins the spinal column (Augustine calls this area a *cervix*), the nerves which then move the consciously controlled body. But it takes a bit of time for this to occur. So the path connecting sensation to movement runs through the middle of the brain, upward from the face into the mid-ventricle and then down again to the spine. This middle chamber processes the sensation-derived intentions in order to make them memorable, for we could not act on our intentions without the aid of memory, since we have to remember what we perceived (in the anterior chamber) in order to be able to move about in response (from its stimulus having reached the posterior chamber). Without memory we would immediately forget every sensation we receive, and so be frozen in time and unable to move.[58]

[57] Augustine discusses at some length the sensation-conceptualizing process, and the role remembering plays in understanding, in *De trinitate* VIII, IX and XI.

[58] *De Gen. ad litt.* VII. 18: "et quoniam corporalis motus, qui sensum sequitur, sine interuallis temporum nullus est, agere autem interualla temporum spontaneo motu nisi per adiutorium memoriae non ualemus, ideo tres tamquam uentriculi cerebri demonstrantur: unus anterior ad faciem, a quo sensus omnis; alter posterior ad ceruicem, a quo motus omnis; tertius inter utrumque, in quo memoriam uigere demonstrant, ne, cum sensum sequitur motus, non conectat homo quod faciendum est, si fuerit quod fecit oblitus." [And since bodily movement, which follows sensation, always involves intervals of time, and since we cannot perform a voluntary movement over intervals of time without the aid of memory, the brain is proven to have three ventricles, one in front at face-level, from which all sensation is controlled; a second in the rear at the neck, from which all movement comes; a third between the two, in which they [=the medical writers] show that memory is active; otherwise, since movement follows from sensation, a person could not join

Boethius also analyzed the sensory-cognitive movements of the mind in terms of the physical elements from which humans, like all created things, are composed – earth, water, air, and fire – on an axis of relative thinness and density (*raritas/soliditas*) and relative mobility (on an axis of *mobilitas/immobilitas*), for form never occurs in nature independent of the movements shaping it. As Boethius writes in his *De trinitate*: "Physics deals with motion and is not abstract … for it is concerned with the forms of bodies together with their constituent matter, which forms cannot be separated in reality from their bodies. These bodies are in motion, the earth, for instance, tending downwards, and fire tending upwards, and the form which is joined with the matter takes on its motion."[59]

This Neopythagorean physical analysis of corporeal composition was elaborated by a twelfth-century student of Thierry of Chartres (as he says) and likely also of Hugh of St. Victor, named Clarembaldus (Clarembald; d. 1187), archdeacon of the cathedral in Arras. He composed two commentaries on Boethius's *De trinitate*, in the course of which he characterizes how our material sensations rise through changes in their elementary composition, becoming abstracted to a state of knowability by the human intellect. Corporeal data from the external senses, composed mainly of moist earth, is drawn into the front of the head, called *fantastica*, and fashioned into images. But those data also have a good deal of air in their mixture, and *fantastica* uses this air together with their moisture to fashion and imprint the sense-derived images in the mind. "Insofar as the stuff of air is more refined than in a composition formed by means of the hands or eyes or nerves [which are more earthy], just so cognition using mental imaging is more refined than that which is only sense-derived." Our reasoning power (*vis rationis*) uses these products of *fantasia* in order to understand our world, a fine Victorine-style description of Bilah and Rachel acting together. But for

together what needs to be done, if they had forgotten what happened before]; my translation, consulting that of Hill, 334–5. See also the surrounding sections 12–17, 19–20 of *De Gen. ad litt.* VII. For the names of the *medici* to whom Augustine likely is referring, including the Roman physician Vindicianus, whom he knew, see Taylor's translation (1982), vol. 2, pp. 246–7 n32. The frozen state Augustine describes here is exactly that of the damned souls in *Inferno*: even those appearing to move are not doing so by choice, and their memories are frozen in their pasts, endlessly repeated.

59 Boethius, *De trinitate*, II; LCL 74, pp. 8–9: "Nam cum tres sint speculativae partes, naturalis, in motu inabstracta … (considerat enim corporum formas cum materia, quae a corporibus actu separari non possunt, quae corpora in motu sunt ut cum terra deorsum ignis sursum fertur, habet que motum forma materiae coniuncta)." Boethius himself in this treatise takes a firmly Platonist position on the divine Reality of Forms.

full rational understanding, the soul needs intelligible thought-objects that are much more refined and rarified than the airy but moist sensory-images, and so "in the middle part of the head, which is called *logistica*, that most refined air, which is called in physics 'ethereal light,' is kept enclosed." During this stage, the embryonic concept's physical mixture is altered again, to a more fitting elemental proportion out of which is born (*innascitur* – note the gestational model) another of the soul's abilities, its *vis discretionis*, the ability to judge whether something is good or bad ethically, true universally or only particularly, or is entirely false. And "[b]y the measure that *lux etherea* is more subtle than air, by the same measure the soul with the support of *lux etherea* can discern the more refined reality within the phantasm processed from a material object."[60] This physical account of the conceptualizing and abstracting process required in mental activities, steadily rising as well as de-materializing, gives it a more vertical movement than the more horizontal organization of the Avicennan account (shown in Figure 3).

In Boethian accounts like that of Clarembald, sensory data composed from earth and moisture collects in the "common senses" (*communis sensus*) compartment of the brain, then is recomposed with more water and air by imagination (imagination's products are always watery in these accounts, forming images quickly that soon lose their shapes and can readily lapse from our minds unless we attentively make them recollectable).[61] From the forebrain, imagination's *phantasms* rise to the middle of the head, becoming increasingly airy; here *logistica* (reasoning and recollecting) alters their elemental composition again so that their airiness will illuminate in the fiery light

[60] Clarembald of Arras, *Tractatus super librum Boetii "De Trinitate,"* II. 6–7 (ed. Häring 1965, p. 108): "Quanto igitur aeris substantia compositione manus vel oculi vel nervi subtilior est tanto imaginationis cognitio subtilior est sensuali. Cum vero anima secundum vim rationis rem aliquam comprehendit tanto in sua consideratione fit subtilior atque perspicacior quanto instrumenti sui essentia imaginationis et sensuum instrumenta praeterit subtilitate. Siquidem in media capitis parte, quae *logistica* dicitur, aer subtilissimus qui a physicis *lux etherea* vocatur tenetur inclusus. … Unde vis discretionis … ipsi animae innascitur. Quanto igitur lux etherea aere est subtilior tanto anima lucis ethereae amminiculo subtilius rem per phantasicam a re discernit." My translation; an English translation (2002) of the whole text is available.

[61] In his *De memoria et reminiscentia*, Aristotle emphasizes that whereas the young, being mainly moist and warm of elements, are able to learn quickly they also are most liable to forget, since their images quickly lose their shapes. The old, being cold and dry, are slow to fashion new cognitive images but remember their learning much longer. This biological account of remembering became standard in discussions of *artes memorativae* through the fifteenth century and later – for an influential example, see Aquinas's *Commentary*, trans. in Carruthers and Ziolkowski 2002.

of *lux etherea*, whence (*unde*) the mind's judging and analyzing power (*vis discretionis*) can understand intelligently the real Truth within our particular, human sense-derived cognitions.

Thinking, Meditating, Contemplating: Richard of St. Victor, *De contemplatione*

The term *contemplatio* (verbal form, *contemplor, -ari*, another deponent verb) was said by Varro and Cicero (and in the Oxford Latin Dictionary), to combine the prefix *con* with the noun *templum*, initially a sacred outdoor space, marked off by an augur's staff and sometimes also with trees, for viewing intently what the gods might show.[62] Cicero uses it for the mental attentiveness required for persuasive *evidentia*. For example, in a daring oration (in 45 BCE) addressed to the newly victorious Julius Caesar in Rome and in Caesar's own house, Cicero begged clemency for his conquered counterpart, the Galatian King Deiotarus. Cicero urges Caesar to imagine the conqueror and conquered kings side by side: "Do but picture to yourself these two kings, contemplate in your imagination what you cannot contemplate with your eyes; and assuredly you will concede to your compassion what you did not concede to your wrath."[63] Contemplation required attentive imagining with feeling and desire as well as reason; it was always a careful mental exercise; and it had always some sacred purpose, in this case petitioning for the divine clemency that only kings can grant. Contemplation is part of the rhetorical practices for invention and disposition – it was not exclusively associated with augurs and priests, though it was a particular kind of rhetorical *enargeia*, a "placing before the (mental) eyes" (*ante oculis ponenda*) reserved for the most persuasive of arguments.

62 OLD, s.v. *contemplo, -are*; also Lewis and Short, s.v *contemplor, -ari*. For medieval examples see DMLBS, s.v. *contemplari*. Varro has a lengthy discussion of the word, which he links to Latin *tueri* ("seeing" and "attending") as well as to *templum*, *De lingua latina* vii. 6–11.

63 Cicero, *Pro Rege Deiotaro* 40 (LCL 252, p. 539): "Propone tibi duos reges et id animo contemplare, quod oculis non potes: dabis profecto id misericordiae quod iracundiae denegavisti." Cicero and others associate the action also with the verbs *considero*, "consider," and *intueor*, a synonym of *contemplor* in classical Latin, though in medieval Latin (from Augustine on) *intueor* came to refer to the purely intellectual, illuminated, and immediate kind of knowing that John of Salisbury speaks of in *Metalogicon* (discussed above). See Nash 1969 (on Augustine), A. Kenny 1993 (on Aquinas), Marenbon 1987, and Tachau 2006 for good overviews of the question in later medieval philosophy. Practices of inner seeing in monastic apophatic prayer are well discussed by C. Stewart 2001, 2011; and the essays in Børtnes and Hägg 2006.

Before 1162 (probably), Richard of St. Victor composed his greatly influential treatise *De contemplatione* (its English name, "The Mystical Ark," comes from Richard's use, as his primary invention device, of the imagined architecture of the Ark of the Covenant, one of the biblical *geometricalia*). Richard distinguishes a trinity of cognitive activities involved in human understanding: cogitation, meditation, and contemplation. Cogitation is ordinary thinking; meditation is disciplined study, including *lectio divina*; and "contemplation circles around [*circumvolat*] all things and when it wants to [*voluerit*] it balances itself in the heights," rather as a hawk adjusts its wings and tail to hang motionless in the high air.[64] "Thinking wanders about, reasoning investigates, contemplation marvels."[65] Such exercises in speculation themselves involve three activities: *imaginatio*, *ratio*, *intelligentia*, imagination, reason, intelligence. "Out of [*ex*] imagining comes cogitation, out of reasoning meditation, out of intellecting speculation."[66] In human thinking, these three act in trinitarian synergy, in the manner which Augustine (from whom Richard takes the idea) called the image of God in the human soul. Augustine defined the trinitarian aspects as *memoria*, *voluntas*, *intellectus*. And Richard's earlier Victorine predecessor, Hugh of St. Victor, also employed this by-then traditional Augustinian trio. So the question arises why might Richard wish to change this terminology? And in what ways might his changes be significant?

Richard's three terms are taken from Boethius's *Consolation of Philosophy*, Book V, prose 4, Philosophy's explanation of what John Marenbon, in his essential studies of Boethius's philosophical logic, calls the Modes of Cognition Principle.[67] Things are comprehended by humans, Philosophy says, not according as they actually are in themselves but, rather, *secundum cognoscentium facultatem*, according to the particular ability [*facultas*] or

64 Richard of St. Victor, *De contempl.* pars 1.3 (ed. Grosfillier, i.92; my translation, and see also Zinn, 155–6): "Contemplatio autem omnia circumuolat et cum uoluerit se in summis librat." By Richard's time the analogy with such flight was a common trope, probably influenced by Gregory the Great's *Moralia in Job*, where someone in profound contemplation is likened to an eagle soaring on high (see esp. Bk. 31.48). See also Prov. 23: 5. All my further quotations of *De contemplatione maior* are from Grosfillier's edition; the translations of it are mine. For English readers' convenience I have also cited page references from G. A. Zinn's 1979 translation, which used the older text in the PL. Readers should be aware that Grosfillier's edition differs from the PL on occasion.

65 *Idem*: "In cogitatione euagatio, in meditatione inuestigatio, in contemplatione admiratio. Ex imaginatione cogitatio, ex ratione meditatio, ex intelligentia contemplatio."

66 *Idem*: "Ex imaginatione cogitatio, ex ratione meditatio, ex intelligentia contemplatio."

67 Marenbon 1987, 2003.

power [*virtus*] of each mode of cognitive activity. So the various abilities of the senses, of imagination, reason, and understanding, may each cognize the same individual person [*ipsum hominem*] but will do so in different ways, just as we know a particular object differently when we see it than when we touch it or hear it. Mounting above [*supergressa* – notice the *gradatim*, step-by-step model employed in this verb] and comprehending all the cognitive modes as though in one encyclopedic embrace [*ambitum universitatis*] is the sharply focused gaze of the understanding mind, *pura mentis acies*, a concept notably important as well in Augustine's presentation of human understanding.[68]

There is little if anything that would have been unfamiliar in Richard's terminology. The Boethian turn in twelfth-century thought has long been noted, perhaps most influentially by M. D. Chenu and Pierre Courcelle in their various studies of Boethius, and of the *Consolation of Philosophy* in particular.[69] For the Victorines, credited (by Bernard McGinn and others) with giving a new rational organization to the traditional motifs and procedures described for monastic contemplation by Augustine, Basil, John Cassian, and Gregory the Great, Boethius offered the obvious bridge between Aristotle, Plato, and this monastic tradition. So it is unlikely that any of Richard's fellow Victorines would have found it remarkable to read *imaginatio*, *ratio*, *intelligentia* instead of *memoria*, *voluntas*, *intellectus* in a discussion of the abilities of the mind. *Voluntas* is the motive force of contemplating throughout Richard's analysis. He takes for granted the contemplative's *desire* to reach the heights of prayer, desire that burns more intensely (*vehementer*) as his thinking mounts upward, like a bird rising into the high air.

For Richard, speculation (*speculatio*) invokes two puns: *speculum*, "mirror," which he adopts from II Corinthians, understanding divinity through puzzles and conundrums, darkly as though in a mirror, *per speculum in aenigmate*. And, less commonly (to us), he also understands mental seeing to be as though one were gazing from a watchtower, *specula*.[70] In his works on prayer, Richard focuses particularly on the nature of human thinking, and especially on *where* the thinking mind turns. As Grover Zinn remarks,[71] Richard does not maintain that contemplation proceeds from nature to spirit through completely different sorts of thought activities but, rather, that what the mind thinks *about* changes,

68 See Books IX–XI of Augustine, *De trinitate* especially, but the *acies mentis* (or *acies animi*) idea is found throughout his psychology.

69 Chenu 1968, Courcelle 1963, 1967.

70 On the watch-tower image, see Kupfer 2016, 2017. Richard discusses the pun at some length in *De contempl.* 5, 3.

71 See the Introduction to Zinn 1979 translation, pp. 24, 36–38 esp.

the "scope" (*scopus*, "target") of our speculation moving from natural things and their causes (as philosophers do) to considering divine matters such as the nature of God as the One principle of creation (ideas *supra* reasoning only about Nature, but still humanly intelligible) and then to God as the Trinity, a revealed truth that does not make any humanly rational sense, hence is *praeter* ("beyond") reason: the only two other such revealed matters are the Incarnation and the Eucharist. As the speculated object changes so too do our modes of thinking. It is not the case that all sensory knowledge vanishes, but that the sense-based concepts employed are spiritualized, their phantasms directed toward "spiritual" imagining and knowing rather than nature. Methodical reasoning does not disappear but is directed toward increasingly cosmic and divine matters until, like Dante's *alta fantasia*, it reaches its limits and may become powerless in a God-initiated *raptus*.

It has been an enigmatic fact to me that in his treatise, Richard never abandons his chosen organizing structure, the linguistically, syntactically measured *figura geometrica* of the Exodus Ark (Exod. 25ff), visualized materially as gold and wood, poles and rings, its "mercy-seat" and its gold cherubim. These are throughout the treatise carefully envisioned in detail, and on occasion even touched and heard, though mentally. Clearly for Richard, the differences among stages of contemplation lie with their objects, moving towards an understanding whose goal is no longer just the Ark's material characteristics as described in Exodus, but which seeks to fashion and see them inwardly, intuitively. But in this process, the natural materials do not just disappear, any more than the marble stone vanishes when it is sculpted.

Richard speaks of three "heavens" of thinking about divine subjects, all of which are far removed from ordinary earth. The transition to the spiritual heaven from the natural cosmos happens when our self (not just as an individual sinner but as a soul made in God's image) becomes our main cognitive object, because that inward voyage of discovery is the means for dis-covering as well how a trinitarian God might actually be and work. This is why, in *Piers Plowman*, Ymaginatyf requires Will to consider himself again – not this time to confront his own past sinful life on Middle-Earth, but to introspect his selfhood as a soul made in God's image. As Richard's contemporary, Peter of Celle, wrote, "to know oneself in God and God in Himself is the one great, though not insolvable, question; it is preceded by another quest, to know oneself in oneself."[72] That is a fundamental tenet of monastic practice, the all-important Turn of contemplation (*con-versio*) from things of nature

72 Peter of Celle, *De afflictione et lectione*: "Grandis quidem illa una, sed non insolubilis quaestio … se in Deum et Deum in se quaerere; praecedit quidem alia quaestio se in se quaerere" (ed. Leclercq, p. 238.13–16).

to God in Himself, via introspection and speculation even of other kinds of souls. Richard explicitly advises that the examination be enlarged from just one's own self: "Know your own will," he says, but also "think about the souls of the perfect, and even of the perverse, so that from consideration of opposites there may shine forth what it is expedient to imitate and what to avoid. Pay attention to what you know; pay attention to how much you do *not* know."[73] Arguing all sides – *argumentum in utramque partem, controversia*: the reasoning technique was just as powerful and essential for Richard as it was for Augustine.

And this exercise, which includes but is not limited to examination of one's own conscience, in pastoral literature is also called *memoria*. For example, a confessional manual published in Sicily in 1621 is called *Memoria artificiale di casi di cosciencia*, a method of retrieving and reviewing matters that pertain to conscience (=*casi di cosciencia*). Its artificial mnemonic employs the "way of numbers" (*via di numeri*), the numerically addressed locational system familiar from so many other such texts, including one by Hugh of St. Victor on how to mentally inventory the Psalms so you can quickly find arguments in a debate.[74] And Bernard of Clairvaux speaks of *purganda memoria*, "cleansing memory," in a talk addressed "to [novice] clerics concerning conversion."[75] In this same Augustinian-Cistercian spirit, Langland's Ymaginatyf tells Will to amend his life by remembering it. Conversion is not a matter of erasing one's memory (even if one could do so) but of turning and reorienting it, restoring to immediate view *now* what has been covered over, fallen out of our knowledge-inventory, and forgotten [*oblitus*], rather in the

[73] *De contempl.* 3.13 (ed. Grosfillier, i. 310): "Cognosce tuam ut scias quid corrigere … Cogita etiam animos perfectorum nec non et peruersorum; … ut ex contrariorum consideratione elucescat quid imitari, quid uitare expediat. Attende quid scias, attende quantum nescias."

[74] Translated in Carruthers 2008, appendix A. Monks do not require this system to recite or chant the ritual psalms by heart during liturgy. It is in the unusual, *ad lib* invention situations of composition and debate that such schemes prove their value, as Hugh makes plain. The 1621 manual for Examination of Conscience was composed by Anselmo Faccio, an Augustinian friar (one of the two known surviving copies is in the Newberry Library in Chicago).

[75] Discussed in the longer version of Bernard's "Sermo de conversione ad clericos," par. 28; (SBO, vol. 4, 102–3), in which he focuses on this key question for novice monks: "Quomodo enim a memoria mea excidet vita mea? [How can I excise my life from my memory?]" Though God forgives the guilt, he says, the sinful experience is not cut out of one's memory, but it is "turned" and channeled virtuously during the essential remembering exercise of Examination of Conscience; see Carruthers 1998, pp. 94–9.

way Augustine reorients his life-account in Books 8–9 of his *Confessions*, closely interrogating his own life in order, through introspective reasoning, to find God. Such careful searches through one's own memory stores employ ratiocination as their reliable finding instrument.

In rhetoric, reasoning by analogy becomes also a technique of *Augmentation* or "enlarging," which is a basic invention tool along with its contrary, *Brevity* or compression. One should think of these two opposite movements as acting in tandem, setting up a kind of pulse within a composition; this is what keeps it moving and alive or "vivid," as we also say approvingly. Brevity and Augmentation are contraries but complementarily so, each needing the other, each only measurable in relation to the other. Richard recognizes their fundamental counterpoint when he considers the modes of speculative thought: "Enlarging of the mind [*mentis dilatatio*] is when the mind's sharp gaze is more broadly widened and more intensely sharpened, yet this in no way goes beyond the mode of human activity. Raising up the mind [*mentis sublevatio*] is when the liveliness of the intellect divinely irradiated transcends the bounds of human activity, but it does not pass into strangeness of mind [*mentis alienatio*].... Strangeness of mind is when the memory of present things is forgotten ... and transfigured in the divine and, like a kind of pilgrim, the mind transits to a state both alien and inaccessible to human activity."[76] Note that the first mode, *dilatatio mentis*, is entirely human: "in no way goes beyond the mode of human activity." In the second mode, *sublevatio*, one can see the unseen things above but indirectly, translated into familiar analogies. This is the mode of vision that Dante uses (and Hildegard of Bingen, Julian of Norwich, together with all those who describe their visions for others to understand). The third mode is that of Pauline rapture and of the end of *Paradiso*, where "a l'alta fantasia qui mancò possa," imagination, memory, and reason all failing together.

Dilatatio is another term from rhetoric, a synonym of Augmentation. Richard defines it as a simultaneous mental expansion and contraction: "when the mind's gaze [using the Augustinian term *acies animi*] is more broadly widened and more intensely sharpened." A paradoxical motion, to be sure, but one that captures exactly the tandem movements of augmentation and brevity so prized in composing. Inventing produces an abundance of material where

[76] *De contempl.* 5.2 (Zinn, p. 310): "Mentis dilatatio est quando animi acies latius expanditur et uehementius acuitur, modum tamen humane industrie nullatenus supergreditur. Mentis subleuatio est quando intelligentie uiuacitas diuinitus irradiata humane industrie metas transcendit, nec tamen in mentis alienationem transit ... Mentis alienatio est quando presentium memoria menti excidit, et in peregrinum quemdam et humane industrie inuium animi statum diuine operationis transfiguratione transit."

contemplatives "may spread out the sails of their speculation and practice navigating this [mode of] disputation." As "fishers of men," they are "spreading out their net for a catch. Thus, sending out the net now to the left, now to the right of their ship … they often catch an abundant multitude of fish, catching the wandering senses and the slippery affections of men and drawing them out onto dry land."[77] This rational crafting involves similitudes and contraries both. It begins by considering the corporeal things it knows and whose likenesses it recalls, but then "[n]ot being satisfied with what it gathers from likeness, but proving other matters by arguing from differences and connecting them by reasoning [*ratiocinando colligens*] it leaves worldly likeness far behind itself by following out the linked matters [*consequentiarum exsecutione*] … In this way our understanding can rise by means of those things which we know well to those things which we do not know through our experience."[78] Investigating by following a chain of analogies (by their differences and similarities and neighborliness to each other), starting from what we know to discover what we do not know, is a restatement of the textbook definition of *ratiocinatio*.

I emphasize Richard's foregrounding of *ratiocinatio* as basic to even the highest stages of speculative thinking, indeed to everything except direct divine revelation, which is beyond anyone's understanding except from God's gift alone. *Excessus mentis*, *alienatio mentis* (making the mind strange to itself) is well named, for it is the ultimate disruption of any kind of human reasoning method. Richard does not himself claim ever to have been graced with this state, and such intellectual humility is traditional, at least among the mystics of the twelfth century. Among the Victorines a critical difference between *dilatatio* and *sublevatio mentis*, on the one hand, and true estrangement of mind (ecstasy) lies exactly in the mystic's inability to remember ecstatic experience in any detail or to speak directly of it in any comprehensible human language. Richard believes in the power of human reasoning: in this respect he frequents

[77] *Ibid.* 2.8 (Zinn, pp. 184–5): "Qui sibi ad secundam contemplationis huius speciem iam profecit, in tanta materie prolixitate copiose satis inuenit, ubi speculationis sue uela pandat, ubi disputationis sue nauigia exerceat.… Sed qui inter eos sunt piscatores hominum ueniunt sane ad piscandum laxantes retia sua in capturam. Mittentes ergo nunc ad sinistram, nunc in dextram nauigii rete ad Domini preceptum, includunt sepe copiosam multitudinem piscium, uagos scilicet hominum sensus affectus que lubricos concludentes et extrahentes in siccum."

[78] *Ibid.* 2.12 (Zinn, pp. 190–1): "Verumtamen se latius expandens inferiorum angustias intra se includit, nec his contenta que ex similitudine collegit, sed alia ex aliis argumentando conuincens et ratiocinando colligens, consequentiarum exsecutione omnem corpoream similitudinem longe post se relinquit.… Ab ipsis ergo que ad illa uicinius accedunt et que inuisibilium imaginem euidentiorem gerunt, debemus utique similitudinem trahere, ut ad ea que per experientiam non nouimus, per ea que cognouimus intelligentia nostra possit ascendere."

and indeed is himself a master in the *officina Aristotelis*. I want to emphasize this, for characterizations of spirituality have too frequently insisted on irrationality, miracle, and trance as its common and even definitive features. This is simply not true of the Victorines. Both Hugh and Richard certainly describe visions, but they do not claim that what they write down truly describes their rapture. How could they? When experiencing *raptus* one cannot think, nor remember what happened when one is gathered into the innermost Holy of Holies, focused in intense contemplation, and "secluded in a veil of forgetting and strangeness from remembering external things."[79]

Other reasoning methods besides step-like analogy are considered favorably by the Victorines. In the same passage in which Richard likens invention to deep-sea fishing, he says that the fishermen use different nets as they cast widely on every side (a metaphor for the arguments *in utramque partem* that are the method of *argumentum fidei*): "Now they spread out the net of argumentation, now that of exhortation; sometimes to show something true, sometimes in order to condemn falsehood; sometimes in order to elicit something hidden; sometimes to persuade to something just; sometimes to dissuade from a wrong."[80] But Richard mainly employs the technique of constructing *controversia*, pairs of terms compared and contrasted, brought into neighborly conjunction, their similarities and differences enlarged upon, illuminated, and articulated through this reasoning technique. All the Victorines considered speculation to be an art, for it is first necessary to acquire the craft (*techne*) through expert teaching; second, to exercise it, putting to use what we have learned; third to practice it often and with great vigor on a variety of matters. So by stages "the pocket-fold (*sinus*) of the mind is enlarged and made more capacious for all learning and teachings."[81] Richard's analogy conceives of the mind as though housed within a curving fold (the skull), like that formed by draping a Roman toga (and called a *sinus*).[82] One can expand its folds to

79 *Ibid.* 5.2 (Zinn, p. 312): "In tertio gradu arca in sancta sanctorum infertur, et quasi intra uelum collocatur, quando contemplantis acumen ad intimum mentis sinum colligitur et ab exteriorum memoria obliuionis et alienationis uelo secluditur."

80 *Ibid.* 2.8 (Zinn, p. 185): "Nunc ergo argumentationum, nunc exhortationum retia laxant, aliquando ad probandum aliquod uerum, aliquando ad inprobandum falsum, aliquando ad eliciendum aliquod occultum, aliquando ad persuadendum iustum, aliquando ad dissuadendum aliquod iniustum."

81 *Ibid.* 5.3 (Zinn p. 313): "[U]t dictum est gradibus sinus mentis dilatatur, et ad omnem eruditionem uel disciplinam capacior efficitur."

82 The metaphor is used by Gregory the Great: see Grosfillier's edition, p. 590, s.v. *Dilatatio, exiguitas, anxietas*, who cites Gregory's commentary on Ezekiel, II.5: 17. Grosfillier also rightly cites Dante as an influential user of these three related concepts, as of Richard's thought generally. Richard also speaks of meditation

hold more and more knowledge as one continues to learn. It is tempting to link this analogy to the general idea of meditation as starting out with learning that one has previously in-folded in one's mind (*breviter* and in *sedes*) and then can explicate and enlarge in later thinking. Such motion is the basic dynamic pulse of all creativity, all *poiesis*, making, and fashioning. In this analogous way, human fashioning, properly channeled, partakes in the essential folding and unfolding motions of *opifex Deus*, the actions of physical Creation itself.

Above all, Richard stresses the sheer delight and joy of speculative thinking, the "great dancing of the heart" (*magnum cordis tripudium*) as he calls it.[83] He was also a fine watcher of birds, and not just of those suspended over the highest peaks. "The inventing mind of a contemplative goes out and comes back with marvelous quickness (*mira agilitate*), sometimes away, sometimes returning, sometimes wheeling, sometimes concentrated in one place."[84] "We see analogous activity daily," he says, "in the flocking of certain birds. Some rise up, others descend, repeating these motions many times. Some fly fast to the front, others in haste to the rear, and then they meet again, going out and coming back. Some turn to the side, to the right, to the left, repeating many times with great persistence the same changes of their movements … It is possible to see some curve in a gyre (*in girum*), at times a bit wider, at others a bit smaller, yet always returning to the same place. Others suspend themselves with rapid wing beats and by their agitated motion make themselves almost motionless. They do not leave the place where they are suspended, for a long time clumping together within [the dense flock], as though by steadfastly performing this work they exclaimed in human words: 'It is good for us to be here.'"[85]

---

expanding from *sinus mentis* in I.3: "Per intelligentiam siquidem sinus mentis in immensum expanditur, et contemplantis animi acies acuitur [Through our intelligence therefore our mind's pocket is expanded to immense size, and the focused gaze of the contemplating mind made sharper]." These opposite but complementary movements are basic to Richard's analysis of contemplative method (as they were also for Augustine).

83 *De contempl.* 5.18 (Zinn pp. 340–2). Richard often speaks of mental elevation as a kind of dancing and joyful psalmistry, mind and heart "lifted up in a dance of joys" (*gaudiorum tripudio subleuari*), as he also writes in this section, which concludes *De contemplatione*. See also 1.6, describing how the soul in contemplation "dances," *tripudiat*, in its exaltation, as David danced before the Ark of the Lord (2 Sam. 6); and in 5.14, he describes fish when jumping into the air as dancing in a rhythm of joy.

84 *Ibid.* 1.5 (see Zinn, 158): "Nam uiuacitas illa intelligentie in contemplantis animo mira agilitate modo it atque redit, modo se quasi in girum flectit, modo autem se quasi ad unum colligit, et uelud inmobiliter figit."

85 *Ibid.* 1.5 (see Zinn, 158–9): "Huius sane rei formam, si recte perpendimus, in celi uolatilibus cotidie uidemus. Videas alia nunc se ad altiora attollere, nunc se

This remarkable flocking of thoughts in an actively inventing human mind, creating through analogies, similitudes, contraries, bending, wheeling, agitated, suspended, flying up and then down, left, right, out of the flock, then back to it again like the evening murmurings of a flock of starlings prior to settling into their roost, the divinely ordered natural event that Richard possibly had in mind. This is also human Invention, playing by working, working by playing, both at the same time. The dynamic of a flock in action makes the point as no single definition could: *imaginari* and *ratiocinari* (another deponent verb) together conceive, shape, and give birth to understanding. And of course the birds' eventual roost is their bedchamber, their *cubiculum*.[86]

All animals, even fish, can teach us how to joy in creation. Writing of the intense (*uehemens*), immeasurable joy which the soul feels when rising from ordinary human thought, then is elevated beyond thought altogether when it finds itself suspended in the highest heights, Richard says that analogous behaviors can be observed commonly in the habits of various animals.[87] Many

---

in inferiora demergere, et eosdem ascensionis descensionis ue sue modos sepius repetere. Videas alia nunc dextrorsum nunc sinistrorsum diuertere, et nunc in hanc nunc uero in illam partem declinando in anteriora parum uel pene nichil se promouere et easdem discursuum suorum uicissitudines multa instantia multipliciter reiterare.... Videre licet alia quomodo se in girum flectunt, et quam subito uel quam sepe eosdem uel similes, nunc autem paulo latiores, nunc paulo contractiores circuitus repetunt, et semper in idipsum redeunt. Videre licet alia quomodo tremulis alis sepe que reuerberatis se in uno eodem que loco diutius suspendunt et mobili se agitatione quasi inmobiliter figunt, et ab eodem suspensionis sue loco diu multum que herentia penitus non recedunt, ac si operis et instantie sue executione prorsus uideantur exclamare et dicere: 'Bonum est nos hic esse' [Mt. 17: 4, the words of Peter to Jesus at the Transfiguration]."

86 Richard's analogy became famous in writing about contemplation; it is quoted at length by Thomas Aquinas in his discussion of human contemplation, *Summa Theologiae* II–IIae, Q 180, a. 6, arg. 3. In his subsequent analysis, Aquinas makes clear that he does not agree with Richard's truth claims for his catalogue of avian movements, arguing that while human reasoning assesses and articulates ideas using linear or oblique motions it cannot by its nature rise high enough to achieve full circularity, for that is reserved to the Angels' greater purity of intellectual Being, while stillness (*immobiliter*) is the nature of Divinity only, for the cosmos constantly moves. Therefore, he says, Richard's description of these earthly birds as circling and even staying immobile in one place is misleading. Aquinas overlooks Richard's use of *quasi* ("as though") to modify both of these analogies, but *quasi* is an adverb more commonly in use among twelfth-century Victorines and Cistercians than thirteenth-century Scholastics. See Grosfillier's account of this matter in his edition of *De contempl.*, pp. 137–8 nn. 61–4.

87 *De contempl.* 5.14 (Zinn, 334): "uehemens inmensaque letitia dum supra humanum modum excrescit, hominem supra hominem attollit, et supra humana subleuatum

give joyful jumps (*saltus*) and suspend their bodies for a bit in the air when they play – lambs, for example. And fish "when they play in the waters, leap out above the water and leave the boundaries of their native haunts, while they suspend themselves in empty space at least for a moment. So, without doubt, a holy soul seems exceeding the bounds of its native possibility while it is drawn out from within itself by a kind of internal clapping rhythmic beat of its own dance (*interno quodam tripudii sui applausu*), while it is urged to go beyond itself in alienation of mind, while it is suspended wholly in the heavens while wholly immersed in the angelic pageants." This is contemplation that starts fully within human nature, then is drawn in stages above that nature and far beyond it altogether, but it never is without some grounding in its own nature (its own beat), however far into the angelic pageant its gaze may penetrate for the moment, any more than those lively fish can completely leave their watery element.

---

in sublimibus suspendit.... Solent namque in suis lusibus saltus quosdam dare, et sua corpora quantulumcumque in aera suspendere. Sic sepe et pisces, dum in aquis ludunt, supra aquas exiliunt, et natiue illius habitationis sue terminos excedunt, dum seipsos uel ad modicum per inane suspendunt. Sic procul dubio anima sancta, dum interno quodam tripudii sui applausu a semetipsa excutitur, dum supra semetipsam ire mentis alienatione [556] urgetur, dum in celestibus tota suspenditur, dum angelicis spectaculis tota inmergitur, natiue possibilitatis terminos supergressa uidetur." The dance (*tripudium*) Richard seems to have in mind is some kind of a leaping dance (*saltus*), whose beat is marked by rhythmic clapping (*applausus*).

3

homo sapiens as *homo mensurans*

The Ethical Need for Shape

The analogy drawn by Richard of St. Victor between human thinking and the flocking of small birds is a particularly rich one and serves well as a motif for the rest of my study of the essential role geometry plays in all structured forms of thought. What the image captures so well is the flexible, provisional *shaping* that our reasoning requires to be at all productive. Characteristic of evil thought is its inability to adjust its shapes. Dante's Satan is frozen in ice because his desire, will, and mind are thoroughly fixed into just one shape stuck in one place, and flapping his wings can take him nowhere. *Inferno* is also a place only of individual souls with no hint of communal concern or harmonious interaction. But, as we just saw, Richard compares inventive, productive human thinking to birds flocking, in and out, up and down, spiraling, flying apart, coming densely together in a cloud, sometimes seeming to be fixed in the air in one place, but then changing their configurations again, until they gather in place together in their one communal roost. Though many different bird species form flocks, the term "murmuring" is used particularly of flocking starlings, a bird whose habits were known throughout Europe from antiquity.[1]

The three modes of human thinking that Richard identified – cogitation, meditation, contemplation – are distinguished by their movements and directions. Cogitation's motions are shapeless and indirect, wandering about here and there without any particular end in mind. This is now called free association, the kind of undirected, imagination-led thinking that Richard associates

1 Names for *sturnus vulgaris* are found in most European and Asian languages cognate with English; see OED, s.vv. *starling* and *stare*. The bird's medieval English name, *steorling*, is not derived from celestial stars, though a literary folk etymology sometimes made the link. A number of photos and videos of starling murmurations can be found online via YouTube and Wikimedia.

with Bilah (Latin *Bala*) in his other work on contemplation, *De duodecim patriarchis* ("On the Twelve Patriachs"); it is sometimes also called *Benjamin minor*. Cogitation can be channeled into productive thinking, and so is not merely to be condemned as a wool-gathering waste of time. The analogy Richard draws with avian flocking clarifies how cogitation is a process of associating and harmonizing our many random thoughts to form organized communal shapes. Cogitation's opposite, contemplation, stretches upward in disciplined stages to illuminate all things in a single ray, its united light spreading out to encompass all that can be understood through human reasoning, and then seeking to understand even beyond these limits. And meditation, which bridges the two extremes, "presses ahead with great perseverance (*industria*) of mind, often through arduous and rough places, to the end of its path."[2] Meditative thinking is characterized by methodical shaping. Its motions are chosen and directed, and though its paths may not always be easy to follow, it is always "employed ardently in a search for truth," for a meditation is an investigation, *investigatio*, a rational tracking of memory-enabled mental traces, in contrast to Bilah-led cogitation, which is an *evagatio*, "rambling," like Petrarch's delayed and wandering efforts to climb Mt. Ventoux.[3] Wonder and ardent desire are meditation's motive energies that should lead such rational, purposed thinking upward in *alta speculatio*, as Nicolas of Cusa calls it in *De ludo globi*, the bowling game (designed with shape and method) with which I began this study. Or as Richard also writes, in meditation we analyze and ruminate, in contemplation we marvel and wonder.[4] In the very highest watchtower (*specula*) of our marveling speculations, "the rational soul especially rejoices and dances" in the light in which it now sees matters that are beyond reason. In this dimensionless state the soul still moves, but to the measures of its fearsomely intense desire, what Richard also called *vehementia caritatis*, the fierce fervor of love.[5]

2 *De contempl.* I, 3: "Meditatio per ardua sepe et aspera ad directionis finem cum magna animi industria nititur." Richard of St. Victor gives a number of similar distinctions for the modes of human thought in I, 3–6. All quotations of Richard's text are from Grosfillier's edition; all translations of it are mine, after consulting Zinn 1979. Readers are again advised that Zinn's translation is from the PL text, from which Grosfillier's more recent edition often differs.

3 *Idem*. I, 4: "prouidus animi obtutus in ueritatis inquisitione uehementer occupatus"; "In cogitatione euagatio, in meditatione inuestgatio, in contemplatione admiratio" (I, 3). I also discussed Richard's characterization of imagination and reasoning (as "Bilah" and "Rachel") and Petrarch's ascent in Chapter 2, above.

4 *Idem*. I, 3: "per mediationem rimamur atque aliter per contemplationem miramur."

5 *Idem*. I, 6: "In hac utique suprema omniumque dignissima contemplationum specula tunc animus ueraciter exultat atque tripudiat." Aquinas comments (affirmatively) on

Starling murmuration applies to flocking behaviors that are both visual and audible in the sibilant moving of the many wings. The English usage ("m*ur*-mura*c*ion of stares") is attested from the mid-fifteenth century, in a listing of phrases used for associations of birds, beasts, and people under the rubric "a Litel boke of doc*tr*ine for ionge gentil me*n*."[6] The English list also includes "a dissimulacion*e* of birdys [=*bridys*, brides, women]" and "a clatrynge of chok-wes" [=choughs], "an eloquence of lawyers," "a rage of teeth." Martianus Capella's Dame Rhetoric explains murmur's role in reading and composing:

> [M]uch practice and effort are required, for which this is certainly observed to be the norm, that we write out those subject-matters which we wish readily to retrieve mentally; and then, if the matters we are studying should be rather long, they will cling within our minds more easily when divided into parts; next it is necessary to attach markers to each matter which we wish especially to retain; nor are these divisions for reading aloud, but are better used for meditating on in a murmur; and it is plain that memory is more fully invigor-ated at night than during the day, when wide-spread silence aids us nor is our attention [=*intentio*] called away from outside by our senses."[7]

Though this passage is most often taken to be referring to memorization by rote, it cannot only be that, for such initial habits are the domain of Grammar (and of all one's pre-school learning of the alphabet, of counting, basic prayers, and the like) and not of Rhetoric, who is speaking here. To inventory one's own reading usefully, as she describes it, one first selects those matters one may wish particularly to include in one's compositions. (It is wise to heed the common reading advice given in various versions to all students: for

whether the soul feels delight during contemplation in *Summa Theologiae* II–IIae, Q. 180, art. 7. See also Richard's *De quattuor gradibus violentia caritatis*, where he characterizes "vehementia dilectionis quae mentem quiescere non sinit [the fervor of delight that will not allow the mind to be still]" (PL196.1214; my translation). I discuss the necessary role of mental Disquiet in all inventive thought in Chapter 1, above; for a secular use of creative *vehemens*, see esp. n36 of that chapter.

6 Bodley MS. Rawlinson D.328, fol. 171, edited and discussed in H. E. Allen 1936, at 603. Applied to humans, Middle English *murmuracion* is grumbling and "gruc-chynge," the equivalent of modern British "whingeing."

7 Martianus Capella, *De nuptiis* V.539: "sed, ut diximus, magnam exercitationem res laboremque conquirit, in qua illud observari compertum est solere, ut scribamus ipsi quae facile volumus retinere; deinde ut, si longiora fuerint, quae sunt edis-cenda, divisa per partes facilius inhaerescant; tum apponere notas rebus singulis oportebit his, quae volumus maxime retinere; nec voce magna legenda sunt, sed murmure potius meditanda; et nocte magis quam interdiu maturius excitari memo-riam manifestum est, cum et late silentium iuvat, nec foras a sensibus avocatur intentio." Similar advice is standard in all ancient and medieval language teaching.

example, from Hugh of St. Victor: "Learn everything; you will see later that nothing is superfluous." And from the prayer manual attributed to Peter the Chanter (d. 1197): "if you are tenacious of memory, not skimming all over the place, you will find (*invenies*) what you can learn, and can better retain it and always be able to produce it."[8] You may also write out especially apt citations, passages, and so forth, those matters you want to "readily retain (*facile retinere*)" but not for passing school examinations. Rhetoric's advice is directed to people drafting their own works, not to rote-retaining pieces of others' texts. This becomes clear in the last advice given, to meditate on one's selected and inventoried learning using the subvocal murmur of composing, and to avoid at this stage declaiming anything aloud as though your work were finished and ready for its public performance, its Delivery. Composing is always a private activity, best done alone silently (though not always soundlessly) at night, without distractions. Centuries later, Thomas Aquinas still taught that meditation and murmur are the same activity. Lecturing on Psalm 17: 7, "I cried unto my God" (*ad Deum meum clamavi*), he cites Jerome in order to gloss *clamavi* in this verse not as a cry voiced aloud but as an inner voicing: "understand 'clamor' as 'murmur,' something which I considered [while] mulling it over": it is a synonym for meditation, he adds.[9]

Murmur is not-yet-formed speaking, inventing but not finished, the subvocal expression of some germinal thinking in half-formed phrases muttered to oneself, and still a long distance from the fully voiced *viva voce* that brings a work into full being. It is work in process, *tractanda* but not yet *tractatus*, to use a medieval distinction. The ephemeral nature of *murmur* and its media, which are wax or sand surfaces, left-over scraps of papyrus or linen or vellum or, much later on, cheap paper, is the key to its practical and social value: it is entirely erasable and fungible. Murmur serves the purposes of drafting and fashioning as people invent their compositions, and as such it is as much about forgetting and canceling as it is about remembering and conceiving.[10]

But, precisely because they are unformed, hasty, and momentary, and thus are poorly thought out, any public performance of words that are still only at

8 Hugh of St. Victor, *Didascalicon* 6.3: "omnia disce, uidebis postea nihil esse superfluum." Hugh is alluding to advice in Jerome's Letter 22 (to Eustochium), par. 17. Peter the Chanter, *De oratione et speciebus illius*: "si es tenacis memorie et non labilis ubique, invenies quod possis discere et melior effici et semper proficere" (*The Christian at Prayer*, p. 173). Speed reading was entirely condemned as a form of Sloth in this culture.

9 Aquinas, *In Psalmos reportatio* 5.1: "clamor meus in conspectu ejus: Hieronymus: intellige murmur meum, quod cogitavi proponendum: et consonat illi translationi quae dicit meditationem."

10 See Chapter 1, above.

a murmuring stage is dangerous – hence the strongly negative connotation of unshaped murmurs when sounded publicly at full volume, *viva voce*. This is the danger of the mob (and its modern equivalent, internet trolling, which also uses ephemeral, erasable tools), whose mutters and murmurs, in literary convention as in fact, are ever noisy. Complaining and gossiping also are judged as inherently dangerous within communities for similar reasons, for they are types of concealed speaking, muttered or whispered in small private groups, behind the backs of excluded others. The social dangers of murmuring expressed on such misshapen occasions are captured well by Hildegard of Bingen, who speaks of *murmuratio* negatively in terms of its lack of coherent structure: "[A] person should bear their human nature and all their tasks without mutterings, that is without the distorted shapes [=*gibbis*] of sins, and also without hesitations, having indeed complete faith."[11]

There is a long-standing exegetical tradition that, while Adam spoke in a full voice (*viva voce*) in Eden (for example, when naming the animals, which he did aloud with proper nominal syntax), Eve and the serpent both spoke in a *murmur*, not only softly but, more significantly, seeking to persuade another creature secretly and *without any witnesses*. Their murmuring together should have raised a serious warning to Adam, who had spoken *viva voce* until he willingly agreed to Eve's privately offered invitation, and then secreted himself (*abscondit se*) from God, as recounted in Genesis 3: 1–10. As in all rhetoric, an occasion's virtue (persuasive power) depends on the three constantly interacting elements – speaker, speaking, and an engaged, witnessing, judging audience: these should make up every true occasion of deliberative, rational speaking, even in pre-lapsarian Eden. Without the presence of all three, trouble is likely for a community. Like all human linguistic tools, murmur can be used well or badly; how you use it is the key to murmuring's value for any given purpose.

Geometry as Motion

The geometrical figures (*geometricalia*) are conceived fundamentally in the Middle Ages as the traces (*vestigia*) of particular movements, motion defined as sequential changes occurring through time. Such traces can be considered mathematically (according to Augustine, Boethius, and a cluster of learned eleventh-century mathematicians) as *punctus*, *linea*, *superficies*, and *soliditas*,

11 Hildegard of Bingen, *Liber diuinorum operum* I, visio 2.14 (my trans.): "ideo homo humanam naturam ac omnia opera sua sine murmurationibus, id est sine gibbis peccatorum, ac sine hesitationibus, scilicet fidem perfectam habens, portet."

a point, a line, their "covering" or surface area, and their solid form.[12] A point has no dimension, no *intervallum* as dimension was also called in Latin, a word which refers to distances in time as well as space. A line has but one *intervallum*, its length. Surfaces and solids are made of lines considered either in two dimensions (length and width), thus "flattened out" (*in plano*) as a surface only, Latin *superficies* (another word used is *constratus*, "covering"). Or we can measure forms with their natural solidity (*soliditas*), having three extensions in space (length, width, height) and simultaneously in time, for all created things actually exist in three-dimensional space that we also comprehend temporally, in what is now commonly called space-time.

In his school geometry text, *Ysagoge geometriae* [Elements of geometry], Gerbert of Aurillac discusses how things' surface appearances (by which we measure their area) and solid forms can be envisioned and thus measured using the planar *geometricalia*. Though themselves composed of single lines, whether straight or curved, geometric figures define areas (*constratae*) and also solids (*solidae*). If you multiply the lines contained by a given shape's boundaries, its *termini* ("terminations"), you will measure its surface area. Gerbert's term *constrata* for such area coverings is a form of the verb *consterno, -stratus*, "bestrew, thatch," like the dense blankets made up of multiple straws bound together that form the thatching of a roof.[13] Such thatching of lines, Gerbert says, generates (*nascentur*) the measures of a geometrically defined area. And similarly, if the resultant thatchings are piled up on top of one another within that area, you will be able to understand and measure it as

12 Among the mathematical clerics working in Reims and Liège are Gerbert of Aurillac, who was teaching in Reims after 972 and before 989, and composed his influential geometry texts in those years; Bishop Wazo of Liège (d. 1048), who had studied with Gerbert and wrote on aspects of his geometry; and a generation later still, Franco (d. 1083), a scholar of the cathedral of St. Lambert in Liège, who wrote a treatise on squaring the circle. I discuss these scholars further in Chapter 4.

13 Liddell & Scott, s.v. *consterno* 1. The verb *sterno* means "to cover"; this complex form carries the additional idea of coverings made up from many parts, such as straws and branches, tiles, or paving stones. I think Gerbert's choice of the compound form is deliberate, to convey the notion that geometric area is calculated from linear groupings, as thatching is made from measurable bundles of materials. To calculate the area, one counts up the number of bunches covering it. If in a square, for example, the whole surface area contains sixteen units, this can be considered as 16 or expressed as 4 x 4 (four units extended from each of the four sides), as shown in Drawing 1. The technique derives from Roman surveying practices; see Lugli 1996. On the careful visualizing required for all geometric understanding, see Truitt 2012, McNamee 2016 and 2019. Visualizing is a prominent feature of Greek mathematical understanding, as discussed in Netz 1999, 2009a; see also Lugli 1996.

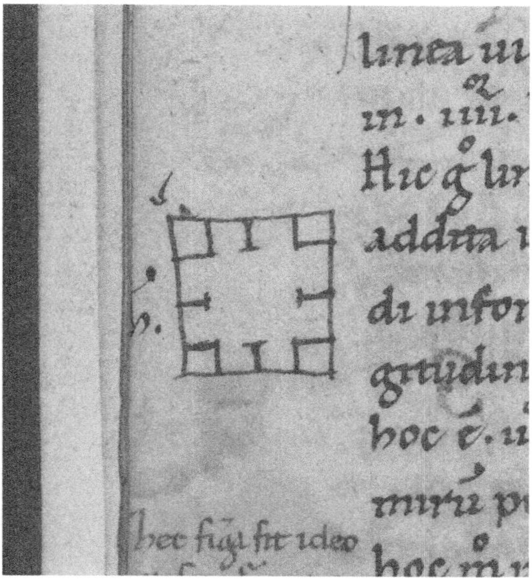

Drawing 1 A sixteen-sectioned square. From a manuscript of Gerbert of Aurillac's *Geometria*, written in Bavaria, mid-twelfth century. Philadelphia, University of Pennsylvania Schoenberg Library, MS LJS 194, fol. 4v (detail). CC-BY-4.0

a solid figure, having height.[14] But clearly, one would pile up such thatchings only mentally, using one's rationally inventive, generative imagination.[15]

An additional way of considering created bodies uses the Platonist dimensional axes of mobility–immobility (*mobilis–immobilis*); sharp–blunt (*acutus–obtusus*, as for the angular arcs of polygons but also applying to ocular focus, sharp or blurred);[16] and thick–thin (*crassus–rarus/tenuis*, referring to

14 Gerbert of Aurillac, *Isagoge geometriae*, PL 139.95: "Aut enim et ipsae lineares, aut constratae, aut solidae intelliguntur. Sciendum autem magnopere est quod per lineares mensuras constratae investigandae sunt. Si enim lineares in se vel inter se multiplicentur, constratae nascentur. Et si constratas itidem per lineares multiplices, solidas invenies"; all translations of Gerbert are mine. McNamee 2016 shows that a plane square shown with diagonal lines drawn from the corners toward the center (but without a medial square actually drawn within it) was conventionally meant to convey height, the equivalent of an embedded square.

15 The practice of measuring in four-fold units of *constrata* lines is at least Roman. The earliest recorded use of the Latin term is in some fragments on practical metrology written by a first-century CE *gromaticus* or land surveyor, named Balbus. He defines linear measures in terms of *pedes*; as used to measure flat (planar) areas of two dimensions, longitude and latitude (such as building sites, fields, etc.), these are named *pedes constrata*.

16 Gerbert also says (PL 139. 93) that "as much as the eye is blunted by the images of natural things, through training such mathematical arts toward spiritual and true matters whatever is to be contemplated appears not a little cleaned up and made sharper [tantis oculum corporearum rerum imaginationibus obtusum per talium

mass). These axes are commonly used in pre-modern Natural Philosophy, which incorporates what we now call Physics but, as we will soon see, is more speculative and cosmologically focused than much of modern physics is comfortable with. In nature, lines also are used to measure time (=motion), the fourth essential dimension of all created objects (whereas I was taught that time considered as a dimension was peculiar to Einsteinian relativity and modern science fiction – but I was also taught that the state of the created universe was basically set in a steady-state).[17]

Ponder this difference: when I first was taught elementary geometry, aged around twelve, I learned that a line was "the shortest distance between two points," the received Euclidean definition. We were then told to take our straight rules and measure said line as fixed in a space from start to end. We were taught that a line was a naturally terminated thing. But many pre-modern Neoplatonist geometries, unlike Euclid (at least in his textbook), defined a line as "unfolding" from a single point: the single point is primary and lines move out from it as time and also space, in the measurable intervals (remember that Latin *intervallum* has both meanings) that fundamentally characterize the created universe. Time is thus a condition of space as space is of time, their imperfect continuity (which is the aesthetic and stylistic principle of *concordia discors*) being the reverse image of divine Eternity.[18] Any particular line *a–b* will measure the same distance (the answer to the question "how long is line *a–b*?" will be the same in numerical units), but instead of thinking of its length only as a fixed end-product one thinks of it also processionally, as a movement continuing through time.[19] This way of conceiving geometric shapes also lies

---

artium exercitia ad spiritalia veraque utcunque contemplanda non modicum purgari et exacui ostendit]."

17 This account is Aristotelian as well, in considering motion and change as measurable in units of time as well as space. Two readable accounts of some modern quantum theories of space-time and the Big Bang, which put Aristotle into the picture as well as Einstein, are Rovelli 2018 (who includes some of Bede's writings about time as well) and Penrose 2010.

18 Quintilian (*Institutio oratoriae* I.x.12) notes another synergy between *logos* and *arithmos*, oratory and mathesis, citing *Timaeus* 34 to define cosmic *armonia* as "dissimilium concordia," which Quintilian says is also a principle of the best constructed arguing.

19 The Aristotelian "species" theory, that light moves from the surfaces of objects to a receiving eye in rays (*radii*) made up of "atoms" of contiguous material particles, each a likeness (*species*) of the object, is a variant of this idea: see Lindberg 1976; Tachau 1988; Akbari 2004. Time also was often conceived as an unfolding sequence of momentary presents – "atoms" or instants; see E. Johnson 2018, describing how Richard Rolle applied this concept of time to his contemplative practice. The fundamental Christian articulation of this idea is Augustine, *Confessions* XI, in which he

within the medieval metaphor of creation as a book (*liber*), such as the "book of the universe," and as a volume, *volumen*. For a *volumen* is not a codex book but a rolled-up scroll, the written units of which one unrolls (*voluere*) as one reads it; the folded-up leaves of a codex must also be unfolded from one to the next in order to read them.[20]

Similarly, a circle is defined as the curving movement traced by the arm of a compass turning around its fixed central point. And a triangle is three lines that join – "kiss" is Alan of Lille's descriptive word – as they unfold and then meet in their angles of convergence, as "right," "obtuse," "acute."[21] The Neopythagorean conception of this figure again emphasizes its motion, lines making angles as they momentarily meet, a dynamic, not a fixed, geometry.[22] The lines we measure with our rules and straight-edges are but records and memories, the vestiges and traces, to use the common medieval idiom, of previous activities now over.[23] The words "circle" and "line" both refer to the motions by which

---

describes how time, via human memory, is made measurable to us, though in itself it is inmensurate (see esp. pars. 30–6). I discuss these matters further in Chapter 4.

20 Diringer 1982, pp. 124–5. The essential difference with Euclid is a difference in the modes of thinking about geometry. Euclidean figures have been abstracted from material creations, and are a matter of rational definitions – their correlation with *Logica* is telling. As *Timaeus* (and Cassiodorus) had made evident, measuring is a requirement of *human* reasoning and can be done only in hindsight – one is measuring movement that is already past. It is a great achievement of seventeenth-century mathematics to measure motions probably and predictively through time; historians refer to this as a process of "algebraicizing" geometry (see Grattan-Guiness 1996). But medieval geometry is pre-algebraic, and respected what were regarded as limits it could not traverse. My particular thanks to John North Crossley for discussion and bibliography concerning this matter.

21 The phrase Alan of Lille uses is *osculo linearum*, angles are formed "by a kiss of lines"; see note 52, below. I discuss Alan's geometrical meditations at length in the next section of this chapter.

22 Albertson 2014, pp. 40ff, gives a good summary of the (complex) thought and equally complex relationships of the mathematical philosophers known as Middle Platonists, centered in the lively Jewish-Greek (and later on, Christian) communities of Alexandria from the second century BCE. This is the mathematical thought which Augustine employs in his *De quantitate anime*.

23 Augustine, *De quantitate anime*, cap. 11, par. 18: "illud enim dico lineae principium a quo incipit longitudo, quod uolo sine ulla longitudine intelligas. nam si longitudinem intelligis, nequaquam profecto intelligis unde ipsa incipit longitudo [I say that the origin of a line where length begins to exist is something I want you to understand as having no length whatever. For, if you take it as having length, you fail completely to understand that a point is where length begins]" (FC 4, trans. McMann). McNamee 2016, 2019 are excellent discussions of this Augustinian-Boethian geometry and its eleventh-century expressions in both words and drawing in manuscripts of Gerbert of Aurillac's *Ysagoge geometriae*.

they were made, *circulus* to the drawing around of the compass line, and *linea* to the cord or thread which measures it once it is terminated, Latin *linea, lineus* (Greek *linon*) deriving ultimately from the name of the flax plant, the source of the threads from which both "linen" cloth and measuring cords are made.[24]

This privileging of the single creating point from which the basic figures all "evolve" (from *ex + volvere*) and "unscroll" as they roll forth is certainly attributable to the Augustinian (Neoplatonist) theological idea that all created forms are "implied" – from *implicare*, infold – within a single, perfect Creator, the single measureless point that initiates the created cosmos that then moves outward, "explicating" (*explicare*, unfold) into the myriad of less perfect reflections that constitute the forms it creates. But it is also tempting to think of the notion as having some practical origin as well as one wholly theological. The inscribed Book of Nature records God's orderly Plan (Logos) in forms that are like letters to be read. Scripts always begin with a single point, as a scribe places a stylus or chisel onto a material surface and then with linear or curving motions unfolds the strokes of each letter as the pen is led around, leaving behind its inked traces. This motion was in fact called *ductus* (the stylus "having been led about"), and learning the *ductus* of the letter-forms particular to various scripts was the basis of scribal training. Actually drawing out the letter forms with one's pen, which as a direct sensory experience is animatedly tactile as well as visual, is the nature of writing in a way that it no longer is when one sets or reads or types with prefabricated letters. The linear and curved strokes intimately and immediately demonstrate how such movements "unfold" in real time from the initial point of one's pen.[25]

24 The ancient relationship of weaving to early Greek mathematical thought (specifically to setting up a warp-weighted loom) has been a focus of study recently; see esp. the work of Harlizius-Klück 2004, 2008, focusing on this relationship; see also Cocker 2017. Weaving is all about forming lines into planned patterns, and is the root conceptual model in the word "text" itself, from Latin *textus* ("textile" has the same derivation). Linen was a common writing surface in antiquity, and is still used in the best-quality papers.

25 The custom of animating scribal letters by turning them into animal shapes or fantastic creatures perhaps owes something to this fundamental quality of scribal writing; see Kendrick 1999. Whatever the complex of explanations offered, medieval figured letters do seem to show a vitality, even to the point of the scribe losing control of the lines, that metal casts and other preset types do not. Perhaps computer graphics can recapture at least an illusion of some of the energy implicit in all medieval lines. As the architect Michael Graves wrote: "Drawings are not just end products: they are part of the thought process of architectural design" (Graves 2012); see also Bork 2011, considering Gothic architecture specifically. Earliest record writing often used squared-off "capital" letters or glyphs, made or painted on hard surfaces; cursive forms written on softer surfaces such as papyrus, linen,

The ordered universe itself comes into being as God draws its existential boundary onto emptiness with his compass, "cum certa lege et gyro vallabat abyssos" (Prov. 8: 27; see also Figure 1). In early medieval geometry manuscripts, the circle figure – always drawn with a compass, its central point often emphasized in ink – is labelled *sphera* (or *spera*, a variant spelling), but never *circulus*. In classical Latin *circulus* is an adjective describing how the noun it modifies is moving. The noun *circulator* denoted the town crier, the official who went about the town proclaiming the news – his official name derives from his movements, what he does, not who he is. Similarly, the figure for a square drawn in a manuscript is regularly labeled *cybus* (or *cubus*, "cube"), and that for a triangle *pyramidis* ("pyramid" and also "prism").[26] Not only should such figures be understood to represent movements but also, in order to be fully comprehended, and thus made useful, they must be understood as signifying a change made in time. The basic figures (square, circle, triangle) are conceived within a solid geometry, not only the two dimensions, *longitudo* and *latitudo*, of a plane. Isidore of Seville defines geometric shapes primarily as solid *figurae*: "Figurae solidae sunt, quae longitudine, latitudine et altitudine continentur, ut est cubus, cuius species quinque in plano."[27] The phrase *in plano* refers only to its appearance (*species*) when a solid shape has been flattened to form a contiguous surface. Another word for such a schematic rendering is *superficies*, the surface appearance that we directly see, which in any solid object does not include the parts that we cannot see, such as the bottom of a cube or the back of a sphere. *In plano* is not an ontological statement about the form itself, but only a statement about how the solid natural form (which has depth as well as length and breadth) *looks to us* when fully flattened onto a flat surface. In the case of a *cubus*, we will observe this as a

---

or parchment make easier the quick, curved lines that give it its name: see Diringer 1982, pp. 53–112 esp. Of course, all written characters begin in a point, whether drawn with chisel, pen, brush, or stylus.

26 Latin *cubitum* (or *cubitus*) is the "elbow": as a linear measure, it is the length from one's elbow to middle finger tip (in English it is also called an "ell"), though it can also be that from nose to middle finger: see the uses recorded in DMLBS, s.v. *cubitum*, OED, s.vv. *cubit* and *ell*.

27 Isidore of Seville, *Etymol*. III.12.1: "Figures are solid which are bounded by length, width, and height, as is a cube, whose appearance is five-fold when flattened." By the late Middle Ages, the Dominican friar Meister Eckhart (Echardus Magister, d. 1328) is definite that two linear coordinates do not constitute a geometric *figura*: "Duae enim lineae figuram non constituunt, ait geometria," *Opus tripartitum: Expositio libri Exodi*, cap. 20, par.132 (commenting on Exod. 20: 5); cited from LLT online.

central quadrate (square or rectangle) surrounded by four others.[28] But this is only how it is drawn in a book.[29] The natural world itself must always be understood and measured in three dimensions (recall the basic connection of *mens* to *mensura*), for material things are all solids and they all, even the most tenuous, have some weight (*pondus*), just as the book of Wisdom says (11: 21). Even apparently non-spatial lines (such as time-lines) have dimension (*mensura*) implicitly, for they are products of a motion with temporal duration which thereby also extends itself as a measurable space.[30]

The dimension of time imparts the essential start of natural creation in this Genesis-based physics. Robert Grosseteste wrote that "The creation of the universe began from nothing and from the initiation of diurnal time." In Genesis 1: 1, Creation is measured first as a Day, and only then articulated and located spatially when God divides the heavens from the earth. "So daily time from which the world began in its first instant, is the measure of the conditional state of the highest heaven and of earth according to an ideal perfection of each kind. And their conditional state was made according to a succession of creations; the Light (=*lux*) creating Day was perfect in the very first instant, and [also] within a day, that is, within diurnal time."[31] Perfect (thus non-temporal) *Lux* expresses itself in creation as *lumen*, the light of diurnal time, from the moment it begins

[28] Vitruvius uses the adjective often to characterize the flat pieces (*tesserae*) used in mosaics and tiling; see *Thesaurus Linguae Latinae* (de Gruyter online database), s.v. *planus*.

[29] Some of the earliest printed books of Euclid are "flaps" (a librarian's term), in which the plane figures can be folded up from the flat page into three-dimensional shapes, an innovation made easier by moveable type and paper leaves but not wholly new; medieval manuscripts containing revolving cut-out circles are not unusual. Presumably these innovations helped to teach the imaginations of the books' student users, as the imaginations of novice monks and apprentice artisans had been trained by their teachers.

[30] See Augustine, *Confessions* XI, 27.33–35, where he examines why humans speak of time (which we experience as sequences of present particular moments) as having "length" and how they can meaningfully measure and compare such apparently non-spatial phenomena. Human memory is the key means. It is typical of how intimate the relationship of time and number was conceived to be that Bede's treatise on natural Time (*De ratione tempororum*) begins with a brief description of the Roman system of finger-calculating then commonly in use; see Wallis 2004, pp. 9–13 esp.

[31] Grosseteste, *Hexaemeron* X.1.4 (ed. Dales, 287): "Cepit igitur mundo creacio ex nichilo et ex diurni temporis inicio … Tempus igitur diurnum, a quo cepit mundus secundum sui instans primum, est mensura condicionis celi supremi et terre secundum perfeccionem specierum suarum. Et fuit conditio sequentium creaturarum successiva, lux tantum faciens diem in eadem instanti primo fuit perfecta et in die, id est, in diurno tempore."

to create the cosmos defined by dimension, number, and weight. Grosseteste adds time to the measures of Creation identified in Wisdom 11. He was certainly not the first to do this, but his definition is notably clear.

This elementary conception of time together with space as the essential character of all natural formation is a strong quality of late antique and medieval Neoplatonist thought, with its roots not only in Plato's *Timaeus* but also in Jewish and Greek-Christian commentary on Genesis. The earliest Latin Hexaemeron was by St. Ambrose, and Augustine's commentaries on Genesis were studied carefully throughout the medieval centuries. Thence it was disseminated in adapted ways among medieval Latin philosophers via a well-traveled route during the early fourth through seventh centuries, including Macrobius's, Martianus Capella's and Boethius's translations and rescriptions of the quadrivial subjects (Arithmetic, Music, Geometry, Astronomy); Cassiodorus's summary of the elementary Liberal Arts; Calcidius's major commentary on the first parts of *Timaeus*; Isidore of Seville (*De natura rerum* as well as his encyclopedic *Etymologiae*); and, summing up for the twelfth century, Bernardus Silvestris's *Cosmographia*.[32]

Geometry was also defined in a rhetorically "brief" form by Cassiodorus (*Inst.* II.6.2): "Geometria est disciplina magnitudinis immobilis et formarum," the study of magnitudes rendered both as artificially immobile (*in plano*) and abstracted from the actual created world. Medieval geometers understood Euclid's theorems and proofs, and the figures that were drawn to demonstrate and apply them, as rational abstractings from the natural forms of particular created things, which – again – by their nature always have some mass (even air and fire have *crassus*, however *rarus* it may be), and thus in nature cannot ever be said to exist in fewer than three spatial dimensions.[33] This is evident from their understanding of *mathesis*, mathematical knowledge. It is Aristotle, at the very beginning of his *Metaphysics*, who set out a three-fold hierarchy of human learning as a process of generalizing and abstracting, starting with *empeira* (particular experiences, Latin *experimenta*), then *techne* (Latin *artes*, the codified principles and methods which enable particular *empeira* to be learned as an orthopraxis teachable to others), and finally to generalized *episteme*, Latin *scientia*, understanding

[32] I have used the translation of Isidore's *De natura rerum* by Wallis and Kendall 2016, and of Calcidius's *Commentary on Timaeus* by McGee 2016; reliable studies of how this difficult material was received in the pre-scholastic medieval centuries include Stock 1972; Wetherbee 1972; Dronke 2008; Fontaine 1983, 2000; Obrist 2004; Mayr-Harting 2007; Albertson 2014; Reydams-Schils 2020.

[33] Even spirits, as created beings, have some *crassus* in medieval theology. G. Smith 2008 gives a delightful account of this matter.

abstracted and universalized in theorems of language and number that can apply in all forms of human sense-derived knowledge.

The thirteenth-century encyclopedist, Vincent of Beauvais, well sums up received medieval opinion (one borne out by modern dictionaries).[34] Latin *mathematica,* he says, is made from two words, the suffix *-icos* which means "knowledge," and *mathesis*, which means "abstract," so *mathematica* is a rational science of abstractions, *scientia de abstractis*, and more succinctly, "Mathesis enim Graecè, est abstractio Latine [*mathesis* in Greek is *abstraction* in Latin]."[35] He distinguishes it from two other modes of philosophical knowing, Metaphysics and Natural Philosophy (Physics), for Physics generalizes its principles from the particular motions and materials of things existing in nature, Metaphysics transcends the particulars of physics to articulate universal principles for those same natural movements, and Mathesis adds another layer of abstracting to the humanly deduced principles of both nature and metaphysics. This fundamental view of all human knowing stresses always that it is partial, contingent, imperfect. God created all natural forms in their real, living beings: to try to understand them, however inadequately, humans invented geometry and the other modes of knowing our world "scientifically," according to our senses and experiences of it.

God created the universe with number, dimension, and mass. But human minds, which are imperfect and limited, require simplified and partial versions of that reality in order to understand it. Euclid addresses Geometry as a system of abstractions closely akin to syllogism. Latin Christian hexaemeral

34 See L&S, s.v. *mathesis*. In his magisterial account of the Galileo controversy, Drake 1980 argued that it was exactly the failure by other astronomers and some Roman clergy to have remembered the orthodox medieval doctrine that all human-derived philosophical principles are reasoned only contingently and imperfectly from nature – and so theology and physics apply to different realms of human knowing and should not try to dictate to one another – that lies at the heart of Galileo's censure by the Inquisition in the early seventeenth century.

35 Vincent of Beauvais, *Speculum maius* (*Speculum doctrinale*), Bk. I.18: "Aut abstrahuntur a materia secundum essentiam, vniuntur tamen ei secundum esse, et de talibus est Mathematica: Dicitur autem a *Mathesis*, quod est abstractio, et *Icos*, quod est scientia, quasi scientia de abstractis. [For they are abstracted from their material being in their essence (=*essentia*), yet made one in their being (=*esse*), and mathematics consists of such things; it is defined as *mathesis* for it is an 'abstraction,' and *icos* for it is knowledge, and so 'the knowledge of abstractions']." *Idem.*, Bk. 16.1: "*Mathesis* enim Graece, est *abstractio* Latine. Fit autem abstractio siue apprehensio quatuor modis. [For the word 'mathesis' from Greek is 'abstraction' in Latin. And it makes an abstraction or apprehension in four ways]," those four being the modes of the quadrivium, four modes of human understanding of physical movement and change in terms of creative rational harmonies.

theology since the time of Ambrose and Augustine focused instead on the processes of physical Creation, how particular material things have come to be, how their bodies were formed and can keep forming new creatures like them. Euclid's geometry was recognized by Macrobius and Martianus (following the fourth-century mathematical writer, Nicomachus) to be a derivative theoretical mode of understanding cosmic measure. Macrobius says that we can comprehend and measure material bodies only from the *superficies* we perceive. But this is not the same as the corporeal thing itself, rather, it is a partial *appearance* of it that our eyes have taken from its actual, particular totality. And it is these apparent boundaries, the *termini*, of the whole creaturely form *as we perceive it* that we measure in order for our minds to comprehend it. For this task, the planar figurations of Geometry are the instruments.[36] In the same tradition, Isidore of Seville defines *Mathematica* succinctly as the human science which considers abstract quantities that have been separated by the human intellect from natural creations solely for its own reasoning requirements.[37]

In this way, Euclidean geometry can also be said to be measuring traces or vestiges abstracted from the boundaries of the solid material bodies that we first perceived. Macrobius particularly distinguishes actual physical bodies from what he calls those "mathematical bodies" which are the concern solely of geometers.[38] It is interesting to consider again from this perspective Villard de Honnecourt's sketches of various sculptured heads and animals he had seen in his travels, which he drew with geometric shapes imposed on them – rectangle, triangle, circle (see Figure 14). Villard (d. 1266) was making a philosophical observation about the generative, dynamic nature of all drawn lines – "li fors des trais de portraiture si con li ars de iometrie les ensaigne [the power of graphic tracings, as these are taught in Geometry]" – rather than

36 Macrobius, *Somnium Scipionis* I.v.6. Macrobius calls this the first stage, *transitus*, the "transition" ("translation" is another name for it) by which the incorporeal intellect is able to understand a material body that it has sensed corporeally by abstracting (note the root *tractare* in this verb) and so drawing an image to "cross over" from one stage or mode to the next in the mentalizing process (from Bilah to Rachel, as it were). Truitt 2012 and McNamee 2016 give persuasive accounts of the common medieval practice of instructing students to visualize mathematical theorems and the stages of logical syllogism. These verbal descriptions are often (though not always) accompanied by a minimal sketch of the figure under discussion, to get imagination started off on the right "track," as it were.

37 Isidore of Seville, *Etymol.* III, praef.: "MATHEMATICA Latine dicitur doctrinalis scientia, quae abstractam considerat quantitatem." This definition is restated many times by Isidore; it was well known.

38 Macrobius, *Somnium Scipionis* I.v.5–8; see also Martianus Capella, *De nuptiis* VII.708–21, on both plane and solid figures.

trying to give practical rules to stone carvers and masons, though he probably had them in mind too. As he writes towards the bottom of this page, as a school subject, Geometry is for contemplating leisurely, *legièrement*.[39]

Humans use different intellectual modes when trying to understand the world, an assumption basic to Aristotelian-Boethian discussions of human knowing, in this case the sense-based mode of natural philosophy (physics) which privileges the particularity of creatures as we perceive and experience them. Our perceptions are the beginning. Gerbert of Aurillac, writing of the geometric figures in the late tenth century, defines their purely mental nature very well. *Punctus* is the origin of all measurement, but in itself has neither measure nor magnitude. *Linea* has length (*longitudo*) but not width (*latitudo*); *superficies* ("surface," "area") has also width, but not height (*altitudo*). Three-dimensionality is conventionally drawn by means of the geometric *figurae solidae*. But these various linear figures do not substantively exist in natural bodies, they are instruments of human creation and understanding.

Sphere, Triangle, Square: Alan of Lille on "The Intelligible Sphere"

The ever-curious and witty mid-twelfth-century Latin poet-scholar Alan of Lille (a figure not unlike Nicolas of Cusa in his own time) is famous as a poet especially for two lengthy allegorical epics presenting a highly Platonized version of Creation, *De planctu naturae* (from about 1160) and *Anticlaudianus*, an account of the building of and journey to Heaven in a chariot (*curriculum*) which, at the wish of Nature, is undertaken by Fronesis (human Prudence or *phronesis*, "wise human judgment,"[40] in order to create and ensoul a New Man through

39 See M. Evans 1980, and more recently Bork 2011 and Murray 2014. As Murray points out, Villard speaks of geometric figures as providing "engiens," "tools" for building and also "ingenuity" in invention, pp. 37–46. One should understand Villard's adverb, *legièrement*, since he sets it in the context of meditation, as monastic (and inventive) *otium* rather than as "leisurely" in any modern sense. Villard wrote in his own Picardian French, yet clearly did not think of geometry as a matter confined to Latinate clergy and the university-taught *artes liberales*. Contemplation was evidently something not utterly foreign to him, nor the use of "li ars de iometrie" in its practices. He provides good evidence that artisanal and clerical learning were not so completely isolated from each other as too many modern histories have assumed.

40 On Alan of Lille's understanding of *prudentia* in *Anticlaudianus* (which, using a Latinized spelling, he calls by the Greek name "Fronesis"), see the note on I.270 in Wetherbee's edition (p. 587). I discussed *phronesis–prudentia* and its intimate connection with educated *memoria* in medieval rhetoric and ethics, in Carruthers 2008. I will refer to Alan of Lille simply as Alan in this section.

an idealized course of education based in the Seven Arts, and culminating in Philosophy and Theology. The *curriculum* itself is built by labors of the Liberal Arts. Reason is its charioteer and the five bodily senses are its horses (their destination is the house of Jove, that is, divine Wisdom approached via Faith and Theology). One could think of it as *regimen principis* of a kind, the Prince in this case being one of the Church, not of a royal court.[41]

Visionary philosophical poetry of this sort is no longer our literary taste, but from antiquity through the twelfth century and well after, it enjoyed considerable influence and many imitators, through such works as Martianus Capella's *Marriage of Philology and Mercury*; Macrobius's prose *Commentary on "The Dream of Scipio,"* which looked back to visionary journeys recounted in Cicero's philosophical musings about Nature and political life; and Plato's dream-filled twin dialogues *Timaeus* (on the gods and creation) and *Critias* (on society). Basic to compositions in this whole genre, classical and medieval, are the twin activities of marshaling one's resources for the journey of composing (Inventing) and simultaneously analyzing its planned stages through a kind of mental geometry, in which the arguments one invents are at the same time placed in relation to one another using structures that can be visualized geometrically, and often architecturally. Above all, Inventing is a *rational* procedure, one that recognizes that we compose and create with structures we have chosen and to an extent pre-planned and visualized, and which we can then use as our tools for additionally developing and refining our inventions.

In addition to the poems just mentioned, Alan also composed a brief prose meditation or *sermo*, as D'Alverny titled it, whose theological theme is stated in geometric terms: "Deus est spaera intelligibilis cuius centrum ubique, circumferentia nusquam [God is humanly comprehensible [as a] sphere whose center is everywhere, circumference nowhere]."[42] Other writers attributed this maxim to the mysterious Hermes Trismegistus, but Alan states firmly that its author is "magnus retor Tullius," the great orator Cicero, "who, just as he outshone all humanity in the beauty of his rhetoric, also gave voice to

41 Alan's Latin poetry has been edited and translated by Wetherbee 2013. All citations of Alan are from this edition. I am much indebted also to Wetherbee's meticulous Introduction and notes. D'Alverny 1965 is essential, for she first recognized and published Alan's *sermo* on the intelligible sphere (it is not in the PL).

42 An alternative reading, "infinite" for "intelligible," is sometimes given for this maxim, but Alan's focus is on how, by using a sphere as a structuring tool, divinity can be made comprehensible by human knowing; it is not on the geometry of the figure per se. This is why I have added "as a" to my translation. Wetherbee, following a suggestion by Peter Dronke, thinks the maxim is most likely Alan's invention; see his Introduction, p. xliv n15.

the profounder language of theology."[43] Alan does indeed cite "Hermes" as an authority elsewhere in his composition (see par. 27), so his attribution to Cicero here is clearly deliberate. And he invokes Cicero (as "Tullius") not as a philosopher and author of works on nature and creation, as medieval writers commonly did, but as a great orator, *magnus retor*. Yet nothing like this maxim occurs in Cicero's works.

So the puzzle is why Alan chose Cicero as his chief authorial source instead of the revered ancient authority on science and physics, Hermes (or "Hermes Mercurius" as Alan later calls him). And specifically, why emphasize Cicero's mastery of rhetoric, of persuasive argument and complex style? What might persuasion – verbal shaping through trope and metaphor – have to do with the mathematical subject of geometry?[44]

The sharp division that is now often assumed between mathematics and language, "numeracy" and "literacy" taught in school as two unrelated conceptual systems, is simply not true in pre-modern education. The building process of education, from sounds to syllables to words to grammar's clauses and sentences, thence to dialectic (basic logic) and rhetoric, merges with the mathematical subjects, from arithmetic to music and the rhythms of poetry, to geometry and astronomy in a process of continual unfolding, of implication and explication, later learning growing upward and expanding from earlier learning, like a tree or like the levels within a tower, to use two common medieval models. In his greatly influential work on physical creation in Genesis, Augustine articulates his meditations about the natural geometry of Creation in sentences of extraordinary balance and complexity, as though to reflect God's natural creations in dimension, number, and density in the elegantly crafted, balanced decorum of his complex Latin sentences. This pairing of natural creation and human creativity is also apparent in Alan's prose about the intelligible sphere, which has baffled and at times irritated scholars trying to parse his thinking in a manner more compatible with modern academic style.

That Alan deliberately adopted a playfully difficult style for his meditation is clear from the fact that he did not always write in this way. His stylistically simpler pastoral works on confession and his model sermons for preachers were much copied, and remained influential into later centuries. But his works on creation were evidently an acquired taste (though certainly known at least by reputation to much later poets, including Jean de Meun and Geoffrey

43 Alan of Lille, *Sermon*, par. 1 (pp. 2–3).

44 Alan does not mention Quintilian, though he could have known parts of *Institutio oratoriae*. Since Quintilian himself is indebted to Cicero's wide-ranging interests in natural science and cosmic dream vision form as well as oratory, Alan most likely did find his inspiration in Cicero, just as he says.

Chaucer). This sermon on the sphere exists now only in a single opening of a single manuscript, and is unfinished.[45] Wetherbee thinks it was a "deliberately provocative address to his students,"[46] certainly not a sermon in any usual sense. I wonder if it was ever delivered publicly, but was always intended as a meditation exercise perhaps enjoyed within a circle of friends.

It is very playful, full of puns, new coinages,[47] inner rhymes, ancient authorities (such as Tullius) invoked incorrectly, and some whose very existence is unclear. Sometimes a well-known authority is invoked only to be obviously mis-summarized. For example, a major incident in Martianus Capella's *Marriage of Philology and Mercury*, one of the best-known works from late antiquity on the liberal arts and natural philosophy, is cited only to be immediately altered without apology, in order to suit Alan's narrative purpose. One might even think of his behavior towards his caricatured authors as a parody of academic citation habits, for such freedoms with one's sources of all kinds is a commonplace feature of medieval verbal creativity, recalling various bits and pieces from all one's learning (including Scripture), rephrased, arranged, and shaped in ever-new ways in the music of oratory. Alan solemnly introduces a rhetorical etymology for the word *diadema* ("diadem"); it is, he says, from *duo demens*, "lacking in two ways," just as a sphere has neither a beginning nor an end.[48] As Wetherbee notes, this etymology is Alan's coinage, and he calls attention to it only to undercut its relevance by stating that the circle-diadem handed out in his version of Martianus's story is actually semi-circular, since God ("Jupiter" as Eternity) gives it to his daughter Choe (an unusual Greek equivalent of Psyche, the human soul), because, although a human soul is immortal, it does have a point of origin, and so fulfills only half of the definition of a divinely perfect circle.

Subtopics branch from his main topics profusely, in near-Boethian fashion. Having defined a sphere, Alan notes that there are really four modes for

45 Paris, BnF MS Lat. 3572, fols. 272v–273; for a description of the *sermo* in its manuscript context, see d'Alverny 1965, pp. 297–306.

46 Wetherbee ed., p. xii.

47 For examples of these, see Wetherbee's explanatory notes. There is also an elegant display of chiasmus in par. 22. G. R. Evans 1983 well characterizes Alan as having a "taste for experimental juxtapositions and new combinations of familiar notions," p. xi.

48 *Demens* here is the present participle of the verb *deminuere*, "diminish," not the adjective *demens*, "demented." Though the phonological pun is technically possible, one must ask why Alan would invoke it in this context, for rhetorical puns, like rhetorical etymologies, are always purposefully constructed to delight and instruct a particular audience.

understanding this "intelligible" sphere (par. 4): *sensilis*, *ymaginabilis*, *rationalis*, *intelligibilis*. Their differences emerge from their common, rhyming adjectival ending. Perceptible, imaginable, rational, and intelligible modes are followed by rhyming opposite logical qualities: *formalis*, *deformis*, *conformis*, *informis*; these could readily form the corners of a Square of Opposition. And there are also the physical axes, *mobilis* – *immobilis*, *instabilis* – *stabilis* (par. 9). The physics seems generated by the linguistic syntax – or perhaps it is the other way around, or even both ways at the same time in a wittily reciprocal motion.

In this way, Alan articulates *through his ornamented style* a basic truth of Christian creation doctrine, that the ordered cosmos is created verbally from Logos (as in John 1) *and also* from Arithmos or mathematically (Wisdom 11: 21).[49] Syntax – words and numbers in productively ordered relationship with one another – is the basic condition of both. Like all figuration, syntactic ordering is an action of creating forms, as Richard of St. Victor makes clear by likening a human mind in meditation to the flocking of birds. Alan's style is often now dismissed as too difficult and clever by half, and we can no longer read it as scientific writing, yet he and his twelfth-century peers took very seriously, *as metaphysical truth*, that an act of in-forming (shaping materials) is also an act of placing and adorning – inventing, organizing, articulating *res* ("stuff") pleasingly (*dulcis*) and fruitfully (*utilis*) with Horatian decorum. Alan articulates his thinking in this work as a set of logical experiments, shifting about verbally in a way that to me recalls the shifting formations of murmuration. Alan's diction moves easily between Logos and Arithmos; the geometric figures are explored ethically and theologically as much as they are mathematically. The *decorum* of his chosen style blurs lines between subject matters in a manner that we now find nearly impossible to follow, and yet it well suits Alan's contemplative purposes.

The geometry that Alan invokes in his meditation elegantly implies this temporal, moving, generative principle. In his understanding, a circle, *circulus*, is not conceptualized as a static figure but as a line gyrating *circulariter*, just as straight lines move out *recte*, "directly, from a single point." A semi-circle starts as a straight line of a certain measure, the *radius*, "ray," which has

[49] See esp. Albertson 2014, pp. 40ff, for a helpful account of how these Neopythagorean ideas emerged in antiquity, to be available to Cicero and Varro, and in greater detail to Christians like Origen, Clement of Alexandria, Marius Victorinus, Augustine, Calcidius, and Boethius. Albertson's account makes clear that the quadrivium was, from antiquity on, as essential a part of the Liberal Arts as the trivium. The culturally perceived split of word from number comes later, and may also account for why modern scholars have such trouble reading works like *Anticlaudianus*, and why it seems so strange now.

been drawn out from a point. The basic definition of a sphere is that all lines within it are of the same length, so to construct one, one places the foot of one's compass instrument on this initiating point (*principium*), sets and fixes its divider (to ensure that the distance between point and line remains exactly equal), and then one unfolds from it, in equidistant measures, a line (circumference) that *circumferens*, is literally "carried around" the center point until it returns to its beginning (an alternative idiom is that the *linea* is *circumducta*, "led around" the central point). The entire concept is modeled as particular movements generated from a point, as also are all forms composed from lines (including triangles, prism/pyramids,[50] circles, globes, cones, squares, cubes, and all polyhedra). Alan cites Boethius's elementary definition of a sphere (correctly this time): "*Spaera est semicirculi circumductio* [a sphere is the leading around of a semicircle]," completing the half by returning the line to the place whence it started.[51]

The intelligible sphere, Alan says (unlike earthly spheres which by definition cannot be squared or triangulated),[52] nonetheless *is* "triangulated," for within its bounds is a peculiar equilateral triangle (actually a prism inside the sphere), that of the Trinity, which indeed measures this very peculiar sphere. This mode of thinking is of course a distinctive method of theological contemplation, the *via negativa* of Christian Neoplatonism, in which the approach to knowing divinity is through understanding the essence of eternal Being as

50 In medieval Latin, *prism* and *pyramidis* (both ultimately borrowed from Greek) are synonymous; see DMLBS, s.v. *prism*. The figures drawn to illustrate *triangulus* often show both a triangular prism (with an equilateral triangle as its base) and what we still recognize as a pyramid, a triangle on a square base (see Figure 2, at the far right among the types of pyramids). Boethius, "On arithmetic" and "On music" also copiously elaborates both triangles and prisms.

51 Boethius, *De institutione arithmetica* II.30: "Est enim circulus posito quodam puncto et alio eminus defixo illius puncti, qui eminus fixus est, aequaliter distans a primo puncto circumductio et ad eundem locum reuersio, unde moueri coeperat. Sphera uero est semicirculi manente diametro circumductio et ad eundem locum reuersio, unde prius coeperat ferri. Vnitas quoque uirtute et potestate ipsa quoque circulus uel sphera est." Augustine gives a similar definition in *De quantitate anime*, cap. 10–11 (par. 16–17), in response to Evodius's inquiry as to which geometric *figura* is the most equal in its dimensions. His teacher, "Augustinus," replies, "the figure whose boundary is everywhere consistent [for *concors*], its equality disturbed by no angle, and from whose center equal lines can be drawn to every part of the boundary," in other words, as Evodius replies, "illam figuram quae una linea in circulum ducta terminatur [the figure which is enclosed by one line drawn in a circle]."

52 par. 24: "nec quadrari nec triangulari possunt"; the verbs are two more of Alan's coinages.

the mirror opposite of its image in changing, temporally created Nature. (As a rational thinking technique, it is reminiscent of the White Queen in *Through the Looking Glass*, who sometimes believed "six impossible things before breakfast." I intend that as a compliment to her Platonist intelligence; one should also recall that "Lewis Carroll" was in life an accomplished mathematician at Christ Church, a geometer in fact, who had also completed his studies for the Anglican priesthood, though he was not finally ordained.)

"In this equilateral triangle," Alan writes, "all the sides will be found to be equal [So far, so familiar], and moreover [Now we go through the looking-glass] all the angles are right angles." "These [three right angles] will be found equal not only to two right angles, but also to one. For the three persons [of the Trinity] are equal with one another, and for this reason can be called 'angles,' in the sense that … angles consist in a touching and a sort of kissing of lines" as, in earthly measures, an angle is formed by a touching of lines.[53] Two different contexts for understanding the term "right" are played with here, *rectus* in its particular geometrical sense as a "right" angle, and also *rectus* in an ethical sense, as "just" and "proper."[54] Alan's meditation is filled with such ethicizing

[53] par. 24: "quasi sicut anguli in contactu et quodam osculo linearum consistunt, sic tres personae trinitate … se quoddammodo osculantur." The *compositionis osculum*, a composing and creating "kiss" is also invoked in par. 11, and cf. Song of Songs 1: 1, "Osculetur me osculo oris sui." Triangles are introduced into hexaemeral meditation by Augustine, *De Gen. ad litt.*, IV.7, where he writes of how a triangle (*trigonum*) "rises in steps" (*gradatim ... surgit*) "from the parts" (*partibus*) of the number six, evidently thinking of an ancient figure for a triangle, which is six stones arranged in three rows, three in a line form the base, above them a line of two, and above that one stone forms the apex. In his Hexaemeral commentary Robert Grosseteste cites this passage in Augustine, describing the figure in detail, as I have just done (*On the Six Days*, IX.2, trans. Martin, pp. 269–70). See also Augustine, *On Genesis*, trans. Hill, p. 244 n10. The ethicizing of geometrical relationships is common in late antique Genesis commentary, Jewish and Christian, Greek and Latin, including that of Augustine and Boethius – the method is precursor to the later, more institutionalized analytical method of up-rising "grades" for Scriptural interpretation (literal, allegorical, moral, anagogical).

[54] Alan is alluding to a "spherical triangle," a triangle whose lines are projected onto the surface of a sphere as three arcs of three great circles, that is, circles whose centers coincide with the center of a sphere (dividing it in halves). The arcs subtending the angles of a triangle thus projected will add up to more than 180 degrees, unlike those of planar triangles. Such spherical projection has been a necessary measuring tool for astronomy, building, and navigation since antiquity (and it still is). According to his student Richer, Gerbert of Aurillac was noted for the elegantly complex spheres that he constructed to instruct his students, and for his mastery of the numerical abacus: see Darlington 1947, Truitt 2012. I discuss this further in my next chapter.

of mathematics, a kind of value-laden ethics of physical nature, but one which is not strictly an allegorical geometry or numerology. Alan never says that his metaphors are the verbal equivalents of the mathematical concepts. His mental playing is not so simple-minded. The adverbs *quasi* and *quoddammodo* ("like," "as though," "sort of," "in a way") get a heavy workout in this composition. His metaphors stretch imagination and reason, as contemplation was said to stretch the mind upward into the heights of thought, the *sublimes*, a mentally levitating sort of playfulness.[55]

Indeed the genre of contemplative commentary on the Hexaemeron, the Six Days of Creation, often features such playfully equivalent imagining, using the tools of geometrical and verbal figuration equally in the human effort to understand something of God, however backwards and reversed, *per speculum in aenigmate*. This is a method with roots deep in Greek philosophy, as in *Timaeus* itself, and in Judeo-Christian theological meditation. St. Macrina (d. 379), the sister of Gregory of Nyssa and Basil the Great, and counted among the Cappadocian Fathers and Mothers,[56] writes in her "Dialogue on the soul and resurrection [*Dialogus de anime et resurrectione*]" that the work of contemplation is greatly aided by considering "*tas geometrikas*" that are measured in Scripture (the Greek word for the *geometricalia* commended in Latin commentaries). These, she says, through their sense-perceptible forms, lead us as though by the hand, step by step, toward matters which are beyond human sense experiences.[57] An ancient commonplace held that the numerical harmonies of music are conjoined naturally with human moral behavior, good or ill.[58] Such geometry-based meditational prayer practices were substantially revived and renewed in the twelfth century by contemplative orders like the

55 "[A] centri motu *tendamus* in circumferentiae quietem," Alan writes at the end, "let us stretch out from the motion of the center to the quiet of the circumference" (par. 29, my italics). Note that this statement reverses the more commonly depicted medieval cosmos of a motionless central Earth with the heavenly bodies moving in increasing speeds outward to the First Mover at the circumference.

56 Pelikan 1993 makes a convincing case for Macrina's importance as a theologian, not only a silent spiritual adept. She is called the Younger Macrina; her younger brother Gregory of Nyssa wrote her biography.

57 This text is in PG 46 (here cited from col. 33C); an English translation is in the Nicene and Post-Nicene Fathers series of Gregory of Nyssa's treatises (ser. 2, vol. 5, 427ff). Macrina's authorship of this *Dialogue* is disputed; until recently it was always attributed to Gregory of Nyssa, but several scholars, including Pelikan, have made a good case for Macrina's authorship. Certainly the Teacher in the dialogue is female. Burrus 2005 helpfully reviews the current state of the authorship matter.

58 A maxim stated in the first sentence of Boethius, *De institutione musica* I, and a classical commonplace.

Carthusians, Victorines, and Cistercians, and for their communities by contemplative abbesses like Hildegard of Bingen and Herrad of Hohenbourg.

Attributing to someone named "Claudian" another maxim which appears to be Alan's own invention, that "God imprinted his triangle on the world" (par. 25), Alan also explicates how a triangle is a proper *imago* of the Trinity, and concludes: "Let us imprint this triangle on our minds" in meditation, to help us avoid vice and cultivate wisdom.[59] Again, the usefulness of contemplating the figure, as Alan understands it here, is not only in understanding physical creation but also human ethics, and it is pastorally presented as an aid to rational thinking (par. 28). Somewhat similarly, the earlier contemplative, Hrabanus Maurus, complied a group of meditation picture-prosae, *De laudibus sanctae crucis* (composed ca. 830), in which he used the six-pebbles configuration for the triangles at each end of the Cross, central to his meditation-picture (Figure 15).

Alan then returns to the possibility of "squaring" the intelligible cosmic sphere, in addition to "triangle-ing" it. The circle of the sphere is in a way "squared" (*quoddammodo* again), but ethically so, since it is a truthful and just (*rectus*) source generating the four cardinal virtues as though forming a square. But at this point, his sermon-meditation peters out, unfinished. And his meditation ends by describing a series of movements from the cosmos in which we live and move toward the apparent stillness at this universe's outermost circumference, the boundless boundary between motion and Eternity. The cosmos, including Earth, consists of the changing movements of creation and destruction, the instability of ever-moving time, the geometry of linear and retrograde and circular motions stretching out infinitely – but not Eternally.

It is a distinctive feature of Christian Neoplatonist physics that Infinity and Eternity are not the same thing. Infinity exists in motions stretching through time from a known beginning ("In principio") even though the end is unknown to us, whereas eternity has no beginning: it is without all motion and thus also without time.[60] They are mirror images, and so the temporal one is reversed,

59 Is this vague attribution another of Alan's teasings? Augustine's use of a triangle (*trigonum*) in his literal commentary on Genesis was widely known and cited throughout the Middle Ages; along with the sphere, it is a primary creation figure in Platonic physics. No candidate for this "Claudian" has been identified.

60 See Boethius, *Consol*. V.pr 6, which, together with Book XI of Augustine's *Confessions*, forms the fundamental articulation of this time-centered distinction between Infinity and Eternity. Grosseteste defines it well also, *Hexaemeron*, I.viii.6: "Illam enim mutacionem que est ex omnino non-ente in ens, non potest alia mutacio precedere, nec instans quod tempus inchoat est continuacio preteriti ad futurum, sed solum futuri inicium. [The change from non-existence to existence cannot be preceded by any other change, and the instant that begins time is not the junction

derivative, and partial, formed from particular *creaturae* placed into a divinely ordained syntax (from Greek *syntaxis*, "ordered arrangement"). Syntax is also a grammatical term in Greek, able to encompass both well-made Greek and Latin sentences (*Logos*) *and* elegant mathematical statements (*Arithmos*). The Book of the Universe has a grammar and tenses, but Eternity, being without time, has neither grammar nor tense nor syntax; it is beyond all language, number, and natural order, being *praeter rationem*, "beyond reason," as Richard of St. Victor says. This is another way in which the orders of numbers and of language support one another in Alan's medieval world, all things, including poetry, being made with dimension, number, and density, including "weighty" sounds, measurable by tone, length, stress, and volume.

Arithmos and Sound

The linking of mathematics with both feeling and ethical behavior, and also with the physical health of body and cosmos, is basic to Neopythagorean philosophy and Platonist thought more generally, but with a particular emphasis on number (*arithmos*) as rational harmony. This strand of ancient thought is given much impetus in Latin medieval theological thought through Cicero's account of "The Dream of Scipio." This work describes the movement of the cosmic spheres as expressing one great harmonious chord; only the Earth is silent or out of tune.[61] The idea that the cosmos is filled (and its nature fulfilled) in harmonious Sound was especially studied in medieval cultures through Boethius, in particular his texts *De arithmetica* and *De institutione musica*, which, along with his translations of Aristotle's introductory logical texts (*Organon*), became basic books of medieval instruction in the *artes liberales* through much of the thirteenth century. The

of a past and a future, but simply the beginning of a future; trans. Martin, 60]"; see also note 31 above. Reydam-Shils 2020 discusses this question as it was taken up in both *Timaeus* and Calcidius. In medieval physics, time and motion are thought to come into being instantly and simultaneously, as the extensive unfolding activity that creates space-time. Eternity is without time or motion, thus in itself boundless, but containing and sustaining that infinite temporal-spatial cosmos it desires to express and chooses to form. Many thanks to Catherine Conybeare for several mind-stretching discussions of these matters in Augustine and others, while walking in South Park, Oxford, during the COVID lockdown spring of 2020. The unnatural nature of that time was well suited this subject-matter.

[61] The final part of Bk. VI of Cicero's *De re publica*, it was greatly augmented as a separate work by Macrobius at the start of the fifth century: medieval writers best knew it through Macrobius.

ideas in these textbooks are also reflected in the specifically literary forms of his *Consolatio Philosophiae*.[62]

The link of sound with human behavior is made in the first sentence of his *Fundamentals of Music*: "Music forms a part of us through nature and can ennoble or debase character."[63] Mathematical harmony demonstrates the ethical and aesthetic dimensions implicit in all number, when considered speculatively and not consigned solely to tasks of counting and calculation, for harmony is proportion and ratio, how (and why) One can generate Many, indeed how any creation occurs at all. In Augustine's words, God is *numerus sine numero*, the originator of all number, but without itself being a number (like the point that has no dimension but from which all dimensions unfold).[64] Proportion is also the principle of rational justice and harmony in societies, families, and individuals. Expressed in sequential numeric ratios among the body's humors, proportion is also the principle of health, bodily and mental, a healthy person being like a well-tuned lute sounding harmoniously, in proper balance.[65] As Boethius further observes, citing *Timaeus*, Plato said "*non frustra*

62 Boethius, *De institutione arithmetica*, trans. Masi 1983; see also his helpful introduction. Scarry 1980 convincingly has analyzed the structure of the *Consolatio* in geometrical-metrical terms. As Masi comments (p. 39), in Boethius's vision of the quadrivium (a name he coined) "[t]he meanings of numbers come not from secret and hidden associations but from carefully determined [rational] relationships … extending gradually in widening circles of application to progressively more extensive meanings until the complete mastery of them leads to the comprehension of the moral order and an all embrasive view of the world."

63 Boethius, *De institutione musica* I. proem: "Musicam naturaliter nobis esse coniunctam et mores vel honestare vel evertere"; ed. G. Friedlein, trans. Bower, p. 1. All further quotations are from this edition and translation. Bower's notes are invaluable.

64 This is an essential principle of Neopythagorean physics, articulated at least in the second century BCE but attributed to Nemesius of Emessa in Cassiodorus's influential commentary on the First Psalm. The Nemesian argument, as stated by Cassiodorus, is as much linguistic as it is strictly arithmetic. The concept "One" by itself is not countable but is the beginning of countable numbers, "monas, licet fons atque initium numeri esse uideatur, ipsa tamen non potest numerus dici ['One' can be seen as the source and beginning of numbering but it is not by itself a number]"; Cassiodorus, *Expositio Psalmorum*, ed. Adriaen, vol. 1, line 432. Grattan-Guiness 1996 notes that, by this definition, the first complete "whole" number is two. Cassiodorus's Psalms commentaries often cite geometrical forms to elucidate the structures of the texts; as he memorably claims in his exposition of Psalm 150, knowledge of all the liberal arts can be exemplified through studying the Psalms.

65 The analogy is an ancient commonplace. I discussed balance as a stylistic as well as medical and ethical principle in Carruthers 2006 and 2013. Kaye 2014 has discussed "balance" persuasively as an economic and commercial principle as well.

[not unreasonably]" that "the soul of the universe was joined together according to musical concord,"[66] a rhetorically playful understatement of the essential rational, creative principle of the cosmos that Boethius described throughout his work.

Geometry is integral to this rational principle. For Boethius, like other philosophers in this tradition, thought of the subject as a whole *mathesis*, which did not arbitrarily separate geometry from arithmetic any more than he considered number to be divided from sound (his treatise on Geometry was often copied with his *Arithmetic*, though it also occurs separately, and in somewhat different versions, which editors now label "Boethian").[67] For music is the sounds of numbers and of geometry too, as those sounds generate shapes with harmonic dimensions within time (the generative *intervalla*, "intervals," that are required for all mensuration). For Boethius, the mathematical disciplines are speculative and theoretical, necessary instruments for human understanding of the whole sounding world and cosmos.

In this physics, all sounds are measured in time three-dimensionally, like every other created thing. As vibrations, they have solidity as well as length and width. This idea was taught first in elementary grammar, the essential platform for all the subjects which grow from it, quadrivium and trivium alike. In his *Fundamentals of Grammar* (*Institutiones grammaticae*), Priscian (a near-contemporary of Boethius, but working in Constantinople), defines language first as sounds, *voces*, and only next defines its letters as the smallest possible divisions of (meaningful) sound, but written for the eye, not only sounded for the ear: "litera est vox, quae scribi potest individua [a letter is a voiced sound which can be written out individually]."[68] Each sound (*vox*) consists of a particular vibrating pulse of the air that has measurable dimensions, as all corporeal things do:

> for if air is a body, then sound, which consists of a pulse of air, is shown to be a body, since it both touches the ear and is divided in three dimensions

[66] *De institutione musica* I.1, trans. Bower, 2: "Hinc etiam internosci potest, quod non frustra a Platone dictum sit, mundi animam musica convenientia fuisse coniunctam," paraphrasing Plato, *Timaeus* 35b.

[67] On the meaning of philosophical *mathesis*, see Cassiodorus, *Variae* I. epist. 54. The word had a secondary application to practices of geomancy and divination (famously condemned as *mathematici* by Augustine), but this difference was well understood in medieval Latin: see, for example, the citations in L&S and DMLBS, s.v. *mathesis*, 1 and 2. In the Greek from which the Latin borrowed, the distinction was clear: the form meaning "divination" was sounded and written with *tau* (τ), but the word meaning the sciences of "mathematics" with *theta* (θ).

[68] Priscian, *Institutiones grammaticae*. I. ii. 3 (ed. Keil).

> which define its particular body, that is in height, width, and length, whence out of the whole [=each sound, *vox*] can also be heard its part. Moreover indeed single syllables have “altitude” in their pitch [=*tenor*], “massiveness” [=*crassitudo*] or “width” [=*latitudo*] of breath [=*spiritus*, for the volume of a particular *vox* is a matter of more or less breath], length [=*longitudo*] in time [=*tempus*, as the duration of a particular sound]. A letter therefore is the marker [=*nota*] for an element [of sound] and also the particular image for a writeable sound [=*vocis literatae*], which is understood from the type and number of a figure’s lines [=the scribal *ductus* of each letter].[69]

Since air, as a physical element, is material in nature, if something that also has mass (a sound) strikes it, that physical act sets up a vibration that pulses through the air until it hits our ears, each pulse having a shape that can be measured in three dimensions. This is not, I think, an idea obvious to most of us now (though it is captured by modern sonograms), but it evidently was to Priscian. And his lengthy Grammar was the standard intermediate language text (studied after the elementary, summary rules of “the Donat”) that preceded Dialectic and Rhetoric in the logical *artes* that constituted the medieval trivium.[70] So his definition of “legible sounds,” each indicated by its proper letter (*litera*), is part of the rote-learned, habit-formed way in which from childhood people who could read and write (*literati*) conceived of sounds and letters. Movement is in the very definition, for *vox/sonus* is fundamentally a particular complex of contingent motions, that, as with all geometric lines,

69 *Ibid.* I. ii. 4: “nam si aer corpus est, et vox, quae ex aere icto constat, corpus esse ostenditur, quippe cum et tangit aurem et tripertito dividitur, quod est suum corporis, hoc est in altitudinem, latitudinem, longitudinem, unde ex omni quoque parte potest audiri. praeterea tamen singulae syllabae altitudinem quidem habent in tenore, crassitudinem vero vel latitudinem in spiritu, longitudinem in tempore. Litera igitur est nota elementi et velut imago quaedam vocis literatae, quae cognoscitur ex qualitate et quantitate figurae linearum”; my translation, consulting that in Copeland and Sluiter 2009, pp. 173–4. On Priscian’s four classifications for *vox*, which are analyzed in the first section of *Institutiones grammaticae* on the axes of literate- illiterate, articulate- inarticulate, see Leach 2007, pp. 24–40 esp.; and Bower (trans.) 1989 and 2001. *Vox literata* is both articulate (as discrete units of sound) and literate (separately writeable), and can be composed rationally with other such *voces*. As Leach notes (p. 26), “there was no sharp distinction between speech and song: both are linguistic vocal performances.”

70 Keep in mind that in pre-modern usage *logica* (and *grammatica*) often included all the language arts, not just formal syllogism (the *Organon*). Standard accounts of late antique and early medieval education include Riché 1978 and 1989, Marrou 1982, Hunt 1980, Reynolds 1996, Law 1997; see also the essays in McKitterick 1990.

scribes will then trace notationally with their styluses, thus making humanly memorable and knowable the initial sonic pulses of air.[71]

But our auditory sensations are not only mathematically rational in nature. As *passiones* they also affect our bodies and thus can have crucial effects on human behavior and judgment. The eyes, Boethius says, perceive objects which the mind then measures and so understands intellectually. The data from all our senses need to be understood by our complement of mental powers, but the sounds captured by human hearing have an additional effect – stirring our feelings, desires, and thus our immediate behaviors to a degree greater than other sensations (not every philosopher agreed with this analysis).

> [T]he sense of hearing is capable of apprehending sounds in such a way that it not only exercises judgement [*arbitrium*] and identifies their differences, but very often actually finds pleasure if the modes [*modus*] are pleasing and ordered, whereas it is vexed if they are disordered and incoherent. From this it follows that, since there happen to be four mathematical disciplines, the other three share with music the task of searching for truth; but music is associated not only with speculation but with morality as well. For nothing is more characteristic of human nature than to be soothed by pleasant modes or disturbed by their opposites.[72]

Sound is an *affectio*, a pulse or blow sensed in the ear from the vibrating air by means of the *nervi*-channeled movements of our human animating spirits, and carried into the brain's *communis sensus* together with sensations

[71] S. Williams (2022) has plausibly argued that several of the decorative line-scrolls drawn by scribes in the lower margins of a large thirteenth-century Oxford scriptorium-made liturgical Psalter now in New College, Oxford (MS 322) may show such sound-pulses unfolding from the mouths of speaking and chanting creatures.

[72] Boethius, *De institutione musica* I.1; trans. Bower, 2: "Idem quoque de ceteris sensibilibus dici potest, maxime que de arbitrio aurium, quarum vis ita sonos captat, ut non modo de his iudicium capiat differentias que cognoscat, verum etiam delectetur saepius, si dulces coaptati que modi sint, angatur vero, si dissipati atque incohaerentes feriant sensum. Unde fit ut, cum sint quattuor matheseos disciplinae, ceterae quidem in investigatione veritatis laborent, musica vero non modo speculationi verum etiam moralitati coniuncta sit." Boethius uses the noun *modus* ("a measure"), which Bower translates as "modes" but notes rightly that it has a range of meanings for Boethius (and later writers), including "tuning" and "temperament" (it is often conflated now with English "mood," though the English word developed from an Anglo-Saxon noun *mōd*, usually translated as the "mind" generally). For Boethius, *modus* is an important philosophical concept, signifying a way of understanding such as the difference between reasoning and desiring: see Marenbon 2003. An excellent essay about similar teaching in rhetoric regarding an orator's varying voice (*vox*) is Schultz 2016.

(*affectiones*) from all the other external senses, which *vis imaginativa* as "fantasy" then shapes into the conceptions that "give birth to" our thoughts and ideas. The corporeal, generative nature of this pre-modern cognitive model is omnipresent in the ways that the activities of conceptualizing are philosophically and medically analyzed. The reciprocity of sensory activities, receiving and shaping simultaneously, is also captured grammatically in the deponent "voicing" of the Latin verbs often used for it, especially after the eleventh century: *imaginari* and (much later, and often with an altered meaning) *phantasticari*, also spelled *fantasticari*.[73] The Latin verb meaning "to measure" is also a deponent: *metior, metiri, mensus est*, a linguistic coincidence that likely helped to support its later association with *mens*.

Throughout his works that discuss the powers of music, including the *Consolatio*, Boethius joins hearing to moral virtue via some common medical ideas about *affectio*. A synonym of *affectio* is *passio*. It is important to understand that medieval Latin *passio* is not the same as the modern concept of "emotion." A clue to this medieval difference lies in the fact that *passio, passiones* had a position in the discourse of grammar, to denote what we now call inflections, those suffixes and prefixes attached to the root of a word that inflect (incorporating the Latin verb *flectare*, "bend") and affect its meaning, that "color" its modes in different contexts and for various purposes, for grammar also has its moods and times/tenses.[74] This was the traditional understanding of how the *voces* of orators (and poets) will generate and inflect emotions in listeners. In his *Commentary* on Donatus's Grammar text, the Carolingian scholar Sedulius Scottus (ninth century) defines all grammatical cases and modes as indicating "diverse inclinations of the mind, showing its various affections. *Modus* is named from *motus*, that is from the movement by which the one [affection] changes into another."[75] In other words, *passio* is not

[73] On the medieval development of deponent verb forms in Latin, see Flobert 1975. As noted above, various mental activities are designated by deponent verbs. A quick word search in the LLT databases for *imaginor, imaginari* indicates 22 uses in the Patristic period (nearly all in Augustine), and 1,281 in the period 753–1500. *Phantasticari* was always a rare verb: see DMLBS, s.v. *fantasticari*. A search in LLT databases shows few uses of *phantasticari* as a verb, nearly all in mystical texts from the fourteenth century and later; before then the root, *phantas-* (*fantas-*) was used mainly as an adjective or (sometimes) a noun, often with negative meaning.

[74] DMLBS, s.v. *passio* 5(e); see also the examples given at 5(c) "sensation," the body's reaction to external stimuli, such as scents, light, and sounds. The Latin phrase *tempus et modus* is commonly used in teaching grammar to categorize forms of speech, as it still is in teaching languages, as "tenses" and "moods."

[75] Sedulius Scottus, *Commentum in Majorem Donatum Grammaticum* sect. 439 (ed. Brearley 1975, p. 109): "diversae inclinationes animi, varios euis affectus demonstrantes. Modi enim dicti sunt a motu, id est a motione, eo quod unus mutatur in

something first generated from inside the body and brain, but something from outside that affects (*affectus*) our material senses by vibrating into them the pulses and particles we understand as sound, light, smell (or that we receive internally as taste and touch). Yet, when actively conceptualizing a sensation we have received, our minds also inflect and affect how we comprehend it, specifically what and how we "feel" about it, a grammatical idea somewhat like the Stoic ethical notion of our body's first (immediate) and second (considered) "motions."[76] It is only the recent, modern separation of reason from emotion that has given rise to our understanding of emotions as wholly internal, irrational, and now even hormonal in origin.[77]

In the discourse of rhetorical Invention, the creative principle of "unfolding" is basic also in the *brevis-dilatatio* movement of composition, the pulse-like motion that drives an argument or a narrative along, as an abbreviated topic is explicated and amplified through various tropes and arguments positioned in an oration *in utramque partem* or in a narrative as its varying path. We find this principle at work everywhere, for example in Augustine's self-debate which I discussed in Chapter 1. Illuminating also is another Priscian/Isidore etymology for *letera* (an early alternate spelling of *litera*), as *leg-iter-a*, "reading path" (from *leg-ere*, "read," plus *iter*, "path, journey").[78]

The creative geometry of unfolding and pulsing is articulated in a common medieval theological/literary trope which visualizes an intensely focused point as the appearance of the divine principle, from eternity to the world and within creation itself. Benedict (and many after him) saw Eternity as a point of brilliant light into which the whole world was gathered up; Julian of Norwich saw all of nature as a round ball which Jesus placed in her hand, "a lytille

---

alterum." The modern word "emotion" expresses the same idea of resulting from a motion, *a motione* as Sedulius said.

76 On the developments of Christian Stoic ethical thought, see Sorabji 2000 and Colish 1990. Studies of the (much misunderstood) ethical notion of *apatheia*, which also clarify the differences of eighteenth-century *emotion* and pre-modern *passio*, are C. Stewart 2001, 2011, and McGuire 1988. The OED entry s.v. *emotion* demonstrates the profound changes since the sixteenth century in how the word has been understood and applied.

77 The literature on this matter is vast, especially since the history of the emotions has recently been again a much-studied topic. On the psychology side, see J. LeDoux 1996 and esp. his cautions and revisions in LeDoux 2012; among the reliable histories, Rosenwein 2006 and 2008. Changes in how mental activity has been modeled over time afford an interesting lens through which to view these matters: see Draaisma 2000.

78 See Priscian, *Institutiones grammaticae* I.ii.4, cited also by Isidore of Seville, *Etymol.* I.3.

thynge, the qwantyte of a haselle nutte [=hazelnut], lyggande in the palme of my hande, and to my vndyrstandynge that, it was as rownde as any balle."[79] Dante imagines Divinity in *Paradiso* 28.16 and 29.9 as a *punto* smaller even than the tiniest of stars, about which the nine orders of angels rapidly circle, the closest to the divine point moving the fastest, the circles farther from it more slowly, in reverse of the apparent rotational speeds of the spheres of the created cosmos, where we humans have reasoned that the outer circles must be rotating more rapidly (because of the great, though yet measurable distances involved) in order to maintain their relative positions as we from our Earth-station perceive them.

Dante's divine *punto* then unfolds into a great Rose whose petals are tiers of heavenly souls ministered to by angel-bees (cantos 30–32). And in it, he sees the souls he has just seen and spoken with in their apparently proper spheres "below." Beatrice appears to be both beside him as she has been all along, and also seated high in the Rose. He has no idea of how he got into the Rose, not even a sense of moving, for throughout Paradise (and unlike Hell and Purgatory) there is neither time nor space. We and Dante are there limited by the bounds of our knowing, by the Arithmos and Logos relative to the speed and nature of Nature's light, *lumen*.[80] Within the originating divine Point in which Dante finds himself at the end of *Paradiso*, time and dimension are no more and what he sees both is and is not there (for truly there is no there there).

Time as Shape: The Notion of "the Fullness of Time"

Like space, time also is rationally measured sequentially, in irreducible units unfolding one after another as "moments" (*intervalla*) each of which is perceived completely, if only instantaneously, in the present. Because these units

79 Benedict's vision of "the whole world gathered up as though in a single ray of light [omnis etiam mundus, uelut sub uno solis radio collectus]" is recounted by Gregory the Great, *Dialogus* II.35 (see also IV.8). Julian of Norwich describes her vision in her *Book of Showings* I. 4.

80 Many late scholastic theologians, including Nicolas of Cusa, addressed the question of whether God could only have created in the way that He did, and concluded that, being Omnipotent (*potentia absoluta*), God certainly could have created otherwise or not at all, but also being creative Love (*misericordia, caritas*), God chose to ordain the universe as we know it (*potentia ordinata*). A notably concise, clear exposition of this idea is Oberman 1965, pp. 30–56, esp. pp. 40–7. Hexaemeral accounts, like that of Grosseteste, commonly distinguish between *lux*, the unique divine light that initiated the cosmos, and *lumen*, the illumination that keeps it going and that enlightens human intelligence; the distinction is made by Augustine.

are linear spatial dimensions (as to be measurable they must be), distances readily can be conceptualized as amounts of time spent getting from *there* (start) to *here* (finish). Medieval journeys and geographical distances were for practical purposes measured in temporal units such as days and hours (for example, a village would be said to be two days' journey distant) rather than in miles, reflecting this still-common slippage between time and space.[81]

The unfolding line of History (for in Judeo-Christian theology history is linear, not cyclical) stretches from past to present and seeks to anticipate future. Human minds comprehend this rationally by considering continuously changing time as a succession of momentary "atoms" just as numbers are rationally measurable in sequential units multiplying out from One (the singularity, Augustine's *numerus sine numero* again). Quoting a contemporary dictum, "tempus a creatura coepit [time begins from creation]," Augustine emphasizes just this point: "indeed if there should exist no movement in a created being, whether spiritual or corporeal, by means of which present, past, and future should succeed one another, time would not exist at all. But a creature cannot be moved if it doesn't exist. Wherefore it is more likely that time begins from a creation than a creation from time."[82] Angels do not experience time, and thus have no memory, but all animals including humans do so, because they exist in and remember time as passing. In this way temporal atoms, humanly understood as sequences of momentary presents, once they are set in motion, create the cosmos by unfolding it as time as well as physical space.[83]

But it is not just time as physical motion that unfolds from Creation, for what is implicated in that creating Point is a specific narrative, a divinely ordained salvific History implicit from the moment of Creation.[84] God's "*Fiat lux*" precipitates cosmic Nature and History simultaneously – there is

81 Examples are given in Le Goff 1980 and Bartlett 2006, ch 6.

82 *De Gen. ad litt.* V.12: "[M]otus enim si nullus esset uel spiritalis uel corporalis creaturae, quo per praesens praeteritis futura succederent, nullum esset tempus omnino. moueri autem creatura non utique posset, si non esset. potius ergo tempus a creatura quam creatura coepit a tempore."

83 Bede, *De temporum ratione*, 3: "Minimum autem omnium, et quod nulla ratione dividi queat tempus, atomum Graece, hoc est, indivisibile sive insectibile, nominant." [The smallest time of all, and one which cannot be divided by any reckoning, they call by the Greek word "atom," that is, "indivisible" or "that which cannot be cut"]; trans. Wallis, p. 15. Bede also argues that such an atom, because it is so small, is more apparent to grammarians than to "calculators," since grammarians perceive a temporal atom as a duration of sound, not as a spatial unit of measure. E. Johnson 2018, esp. pp. 31–6, has an excellent discussion of this idea in later English medieval contemplative teaching, and an extensive bibliography.

84 This is the main argument of Augustine's *The City of God*.

no pre-history in this cosmos, just as there is no matter before God created it.[85] On this account, terms like "the fullness of time" make sense in physics as well as in history. In the early fifteenth century, the French scholar and chancellor of the University of Paris, Jean Gerson (d. 1429), a cleric as busily engaged in Church affairs in his day as Nicolas of Cusa was some fifty years later, composed (1414–17) a *carmen heroicum*, as he termed it, called *Josefina*, to promote renewed devotion to the Holy Family by rehabilitating (as he saw it) the figure of St. Joseph. Joseph was traditionally depicted in apocryphal narratives and in art (for the gospels do not tell his age) as the elderly, frail consort of a robust and much younger Mary. As part of an urgent pastoral program of the Church at this pre-Reformation time, trying to instruct a married laity while itself emphasizing its clerical celibacy, Gerson promoted a devotion to Joseph as a young man, Mary's equal in health, deeply caring of his young family, and thus a suitable role-model for husbands.[86]

Gerson justified his new understanding of Joseph by arguing that an elderly, frail man could never credibly have fathered a child, and so, while not denying the miraculous virginal conception and birth at all (nor Mary's own immaculate conception), in order to avoid arousing scandal among their village neighbors Joseph would have had to be much younger than previous tradition had supposed. As Isabel Iribarren notes, by strict syllogistic logic, Gerson was skating on thin ice in making these arguments, but justified himself by invoking the type of truthful argument defined in Aristotle's *Topica* VI, as one of *probabilitas*, a "probable truth" or enthymeme, the truths of societal and ethical goodness and benefit that are invoked most frequently in rhetorical arguments. Gerson openly invokes the *Topics* in his epic:

85 On this point Christian orthodoxy broke with ancient Platonism, by specifically rejecting a *Timaeus*-derived idea that there was unformed material before the initial divinely commanded cosmic expansion. Plato's account was incompatible with Gen. 1: 1, which states that "*In principio* creavit Deus caelum *et* terram [*In the beginning* God created heaven *and* earth]"; see above for the theological argument.

86 An excellent discussion of the interpretative issues that Gerson engaged theologically and ethically in his exegeses of the stories in Scripture is Iribarren 2020. She particularly analyzes these in the context of academic logics of his time, greatly influenced not only by Aquinas but also by Arabic interpretations that included poetic statement as a form of rational analysis (*rationalis*), not only as a matter of language use (*sermocinalis*) and thus elevated rhetoric as a division of moral philosophy, because it addresses how judgment (*estimatio*) is formed beneficially by means other than strictly formal binary syllogism. See the influential account of this logical development by Black 1990, 2000. Copeland 2014 illuminates these matters in the context of vernacular preaching and the concerns of pastoral care, especially in later medieval England.

> Our meditation rashly affirms nothing/ Concerning things unknown, it will be content only to make use/ humbly of conjecturings by means of topical reasoning./ With study, a pious mind can infer what is not written down from the matters that are written.[87]

As Iribarren says, "Such truths result from an exercise of 'estimation or pious belief' (*existimatio vel pia credulitas*) and escape the principle of bivalence of classical Aristotelian logic."[88] They are not logically necessary, Gerson says, but are allowable to those of pious mind and practical common sense in order to promote devotional good.

Scholars' accounts that stress the logical and academic developments which contextualize Gerson's arguing are certainly correct. But the basic Pauline teaching of history as a pre-ordained narrative unfolding in time seems to me also at work in allowing justifications such as Gerson's to stand without ecclesiastical condemnation. History unfolds outward and upward toward a revelatory "fullness of time," events not simply repeating (as some cyclical theories may hold) but moving towards a more perfect revelation in human understanding as well as towards the completion of natural time, as later matters unfold expansively into a greater "fullness" from earlier ones, historical time imagined as unfolding in a spiral, not a simple line. Gerson also invokes this principle:

> Le Saint Esperit revele aucunes foys a l'Eglise et aux docteurs darreniers aucunes veritez ou exposicions de la saincte Escripture lesquelles il n'a pas revelé a leurs predécesseurs.... Moyse sceut plus que Abraham, les prophetes que Moyse, les apostres que les prophetes. Et les docteurs ont adjousté oultre les apostres. Si pouons dire que ceste verité que nostre Dame ne fut point conceue en pechié originel est de celles qui sont nouvellement revelees ou declairees, tant par miracles qui se lisent comme par plus grant partie de saincte Eglise qui ainsy le tient.
>
> [The Holy Spirit revealed several times to the Church and to later doctors several truths or understandings of holy Scripture which were not revealed to their predecessors.... Moses knew more than Abraham, the prophets more than Moses, the apostles more than the prophets. And the doctors added additional (truths) beyond the apostles. We may say that the truth, that

87 Jean Gerson, *Josephina*: "Nil super ignotis igitur meditatio nostra/ Affirmet temere, sola suffecerit uti/ Conjecturarum topica ratione modeste/ Ex scriptis inferre potest non scripta pio mens/ Cum studio, sic certa fides incerta revelat/ Qualia sunt acta vel quae fieri potuere." Online in LLT as Jean Gerson, *Opera poetica*, opus 138 (=*Josephina*), III distinctio, verse 102, which reprints the *Opera omnia*, ed. Glorieux, vol. 4. My translation, after consulting that of Iribarren 2020, p. 263.

88 Iribarren 2020, p. 260. On the truthfulness of enthymeme, see Burnyeat 1996.

our Lady was not at all conceived in original sin, is among those matters which are newly revealed and declared as much from the miracles which are read as by the greater part of the Church which therefore holds it.][89]

This view of history also justifies the idea of *figura*, so well analyzed by Eric Auerbach (and Dante), that the narrative patterns in earlier events foreshadow those more revelatory ones to come.[90] But it is first of all an assumption about the nature of *time*, not, I think, primarily about patterns in *events* linked by their mimetic narrative components (and certainly not about prophecy as mere prognostication). The idea of "fullness of time" (*plenitudo temporis*, Gal. 4: 4) looks forward into a future whose end is ordained though we cannot yet understand it, and also supports the medieval exegetical idea that past events are at best incomplete analogies of those newly unfolding from them.[91] "Fullness" with respect to human history is synonymous with ideas of completion and perfection, qualities that both enlarge and raise understanding in an upward spiral of knowing as we can now better explicate our understanding of what is implicated in past events. It invokes assumptions about both physics and geometry that are at work within any act of recalling and narrating how past events are continuously moving into the present and into the future. In this way, time does not simply end – rather, it "completes" as it becomes perfected. One should imagine human time as a journey that enfolds and enlarges the stages of its way until it completes like a pilgrimage, rather than emphasizing an unknowable End in which time and space are abruptly exterminated.

This stance toward history also helps account for the simultaneous freedom and respect that medieval scholars gave to the genre of commentary, and why they had no difficulty in greatly expanding their author-texts, in ways that modern textual scholars can find bewilderingly free and unscientific. And they are, at least as we reverently try to consider the past in terms of "what really happened," an impossible goal to achieve, as every practicing historian knows. Cassiodorus seems to have first copiously used the late Latin term, *moderna*, to mean "matters in the present" as contrasted to *antiqua*,

[89] Gerson, sermon on *Tota pulchra es* [on Song of Songs 4: 7], in *Oeuvres complètes*, vol. 7, pt. 2, p. 1076), cited by Iribarren 2020, p. 259 n15 (my translation). On spiralling as the movement of the creating "breath" in medieval art see Palazzo 2020; see also Clucas 2005. I discussed spirals in my introductory chapter.

[90] Auerbach 1984.

[91] The Latin phrase occurs also in Eph. 1: 10. Though the King James Version translation is "the fullness of time," the twentieth-century Revised English translation is "the appointed time" (for Gal. 4: 4) or "the time is ripe" (for Eph. 1: 10), phrases that obscure the medieval concept of *plenitudo* in the way that Gerson develops it.

and he uses it neutrally.[92] A millennium later, in vernacular English, a moral distaste becomes apparent for what Chaucer and others called "newfangelness," mere fashion and fad. But that word was never used as a synonym of "modern," which, in European vernaculars, retained its neutral Latin meaning. The notion of temporal Fullness can give a positive embrace to what is new even in sacred history, indeed to *moderna* as socially and intellectually positive – meaning not just "of the present" but "more complete," a positive change. Medieval commentary generally had no difficulty expanding an older authority into the various urgent concerns of the present. The Latin verb *comminiscor* (whence "commentary" is derived; and notice, it is another deponent verb) is, as Uguccione of Pisa says, a compound made from the preposition *cum* and the verb *memini*, "remember".[93] The commentary genre itself is a form of expansive social memory, by its nature acknowledging that remembering is always within and for the present, including planning now for the future. Nor is it an individual sort of remembering, but a communal one – *cum* – the body of commentators over time together explicating and thus expanding a remembered *fons* ["source," "fountain"] in light of their present understanding and for present audiences.[94]

92 See L&S, s.v. *moderna*. A quick usage count via the LLT database shows no uses of this word before Cassiodorus, only 53 in total through the time of Bede, but 1,742 among the medieval writers before 1500, chiefly in order to distinguish between the *patres antiqui* and the *moderni* philosophers. The Latin adjective is from the root *mod-us*, which as we have seen has to do with "measures," including of time. The interesting history of English *modern* can be read in the OED, s.v. *modern*, from its early use in the fifteenth century through the aesthetic values it acquired in the late nineteenth century especially in reference to art and literature, when the Modern first really became a cultural value. The sweepingly pejorative (and ahistorical) meaning often given now to the adjective *medieval* in contrast to *modern*, is a product of the later twentieth century, especially of popular culture in the 1990s, but also of the academic over-periodization of positivist evolutionary Grand Narrative histories: see OED, s.v. *medieval*. Dinshaw 2012 is a fine study of these twentieth-century narrative phenomena.

93 Uguccione, *Derivationes*, s.v. *Memini* (ed. Cecchini, M77 #28, #30, p. 751). The social and sociable nature of commentary over the centuries is especially well visualized in the complicated page layouts of such manuscripts as the large glossed Bibles made in western academic circles from the twelfth century on.

94 Carruthers 2021. Clearly this attitude to authority as well as to history is very different from some modern western ones, which are confusingly both more reverent of the "authorial intention" of an old text and more dismissive of it as having any bearing on the modern present or the future. On the concepts of late antique and medieval *auctores* and *auctoritas*, see Chenu 1927, Minnis 1988, Pollmann 2014, van Engen 2016.

4

The Instruments of Thinking

Thinking as poeisis

By the mid-eleventh century the cathedral city of Liège was the intellectual capital of the Rhineland, greater than Chartres or any other center of learning in northern Europe. Scholar-monks from elsewhere, including Sedulius Scottus from Ireland, often fleeing marauding raids on their monasteries in vulnerable coastal areas, had come to Lotharingian Liège in the mid-ninth century. By the early eleventh century, despite a rocky century involving various wars and insurrections, fierce ongoing quarrels with the papacy over rights of clerical investiture, and issues of legitimacy among the Rhenish warlords who constituted the Ottonian kingdoms, Liège entered its most illustrious period, especially under a particularly energetic, effective scholar-bishop, Notger (d. 1008), a contemporary of Gerbert of Aurillac (Pope Sylvester II, d. 1003).[1]

The later eleventh to mid-thirteenth centuries were a most productive period not only for scholarship in the textual and logico-literary arts of the trivium but also in the mathematical arts of the quadrivium, including music and geometry. This occurred in monastic, contemplative circles, especially among Carthusians, Cistercians, Victorines, and, as lay religious culture increased

[1] Renardy 1979, with an excellent bibliography. Renardy notes: "c'est surtout en astronomie, en arithmetique et en geométrie que les Liégeois se distinguérent" (314). Accounts of the curricular currents in cathedral schools, including Liège, in the eleventh century, are Jaeger 1994; P. Riché's two accounts (1978, 1989) of western classical education before the twelfth century, and his 1987 biography of Gerbert of Aurillac. W. Stevens 2004 discusses in detail just how much recognizably Euclidean material is in the Ottonian "Boethius" geometries (more than often supposed). On the influence of Gerbert and the various conventional *figurae* used for learning *Geometria* see McNamee 2016, and esp. McNamee 2019, which reproduces several varieties of cubic shapes drawn in ninth- to eleventh-century manuscripts, including some much resembling the modern convention. Fulbert, founder of the cathedral school at Chartres, had studied with Gerbert at Reims, though Fulbert was perhaps less of a mathematician than (for example) Notger at Liège: see Riché 1987.

into the late thirteen- and fourteenth centuries, among the religious solitaries such as anchorites, lay mystics, and among various orders of friars, especially Augustinians and Franciscans, and very often women, in religious orders and not. In this final chapter I will examine a few of the many inventive *formae tractandae* using a great variety of *geometricala*, not just the familiar biblical ones, taught and used mainly in monastic and cathedral circles for productive meditation. My choices are certainly not comprehensive, but I hope they may be fruitful for other scholars in their own studies.

Among the earlier of mid-eleventh-century scholars was Franco, part of a group attracted to his cathedral school by the then-bishop of Liège, Wazo (bishop 1041–48), a worthy successor to Notger as a convenor and supporter of some of the best scholars in Europe, including those who had some knowledge of Arabic mathematical work via Gerbert if not directly from the Arabic texts. Wazo also composed a commentary on Gerbert's abacus treatise (*Regula*) so he was well acquainted with the latest study in the subject.[2]

Franco (fl. 1047–83) composed a treatise *De quadratura circuli* [On Squaring the Circle], which he dedicated to Wazo.[3] Franco's introduction to his discussion of this particular problem is illuminating, for he gives a spirited defense of his geometrical interests on the authority of Scripture, naming in particular those descriptions called "the geometricals" by the younger Macrina,[4] which, in addition to their literal content, served as some essential

2 See Burnett 2010, 2015. Wazo's treatise on Gerbert's *Regula* (for calculating with a type of abacus, using Hindu-Arabic numeration) dates from about 1042, when he was first bishop. Gerbert did his teaching of the *artes liberales* in the monastery of St-Rémi in Reims, starting around 972 through the 980s, until he got caught up in the various political struggles of the time.

3 Gerbert of Aurillac also addressed this Archimedean problem in his *Geometria* (ch. 63; PL 139, col. 141). Gerbert likely learned it (together with the indefinite numerical value of *pi*, π – written in Latin words as *tres quattuordecim* – plus a quantity of Indo-Persian-Arabic mathematics) via Mozarabic sources available to him in Catalonia, where he studied (likely at the monastery of Ripoll) as an *adolescens*, according to the Ottonian historian, Richer. Richer's account of Gerbert's curriculum at Reims is in his *Histories*, Bk. III (ed. and trans. J. Lake, vol. 2, DOML 2011). These sources were far richer than any known at that time in northern Europe alone. For an account of Gerbert's life, including his education and teaching, see Riché 1987, who also discusses how much of the work may confidently be attributed to him; see also McNamee 2019. See also Burnett 1994, on the manuscript evidence for Gerbert's important influence on the teaching of geometry in Chartres and Paris into the thirteenth century.

4 PG 46, col. 33–34C: "tas geometrikas," translated in Latin as "instrumenta geometrica," and by Moore and Wilson as "the methods of geometry." I discussed this treatise in greater detail in my previous chapter.

structuring tools of meditation and commentary for a great many other monastic (and, much later, fraternal and lay) contemplatives. "But some loudly assert that the study of geometrical knowledge leads away from holiness, to live as a stranger (*alienam existere*). Surely those who may think thus at least recall that blessed Moses had the greatest skill in this discipline; they do not at all deny that the Ark of the Flood was measured out in Egyptian cubits; these same people do not refute with counter-arguments that the whole Land of Promise was surveyed in geometrical lines."[5] And after also mentioning among his arguments how Solomon "laid out with appropriate measurements" the entire Temple Mount, he cites Ezekiel's vision of the man "with a face like brass" who with his rod and measuring-line measured out the visionary Citadel. "Since this is so," Franco concludes, "what matter of religion is there to prohibit even priests of the Lord from such study? Isn't Ezekiel a priest? Isn't that man whose appearance is of brass Christ himself, the highest and greatest priest?"[6]

Of particular interest to my discussion in this chapter, is Franco's list of scriptural *geometricalia*, for these are exactly the envisioned, measurable edifices that serve so many other meditations and commentaries. The only commonplace biblical structure missing from Franco's list is the Ark of the

5 My translation, from Franco of Liège, *De quadratura circuli*, Part II prol. (ed. Folkerts and Smeur, p. 67): "Sed forte aliqui dicant geometricalis scientiae curam a sanctitate praesulari alienam existere. Nimirum, qui ita putaverint, hi minime recolunt sanctum Moysen quam maxime huius disciplinae habuisse peritiam; hi permensam diluvii arcam Egyptiis cubitis nequaquam retractant; ipsique non reputant omnem terram repromissionis funiculis geometricalibus distributam" (the surveying of lands north of Judea is recounted in Jos. 18: 8–10). Given the eminence of quadrivial studies at this time in Liège, it seems likely that Franco's opponents, like the "Cornificians" whom John of Salisbury, a century later, rebuked rhetorically as the benighted opponents of liberal studies, serve a useful *topos* of disputation, which requires the disputant forcefully (*forte*) to conduct arguments from both sides (*argumentum in utramque partem*) when constructing *controversiae*, just as Augustine did even for his inner debate (see Chapter 1, above). These dramatically constructed debates tend toward hyperbole, so even when there is independent evidence of actual opposition, it is prudent to take into account the conditions of the literary genre, even (perhaps especially) when the vigour of the opposition is well attested. Truitt 2015, pp. 71–82, has a fine account of how such tales of Gerbert's interactions with devils were fostered by adherents of the imperial power during the bitter investiture controversies of the eleventh through the early twelfth century. The earliest of these anti-papal screeds (1085) dates from the years just after Franco's (and Wazo's) activities in Liège. Gerbert was esteemed before then, including by his contemporary emperors, like Otto III, who in his day still held the power over who became pope.

6 *Idem*, p. 68: "Quod, si ita est, quae religio sit etiam sacerdotes domini a tanto studio pro hibere? Nonne Ezechiel sacerdos? Nonnc vir ille, cuius species aeris [Ez. 40: 3], ipse Christus summus et maximus sacerdos?"

Tabernacle as measured out in Exodus, but any monastic scholar would also have known this complex ekphrastic text (*pictura*), since a geometric rendering *in plano* of the Tabernacle, as at the start of the Codex Amiatinus (copied in England before 716 CE from a manuscript brought from Rome), was a feature of Jewish biblical study well before Christianity, and of both Greek and Latin rhetorical inventing from earliest monastic practices.[7] They are among the basic tools of invention, a set of visualized, actively motile thought-engines, useful for envisioning and shaping all sorts of compositions, including making sermons and various meditations on scriptural texts, monastic life, contemplative prayer, theology, and pastoral care. Such composition plans were not limited to glossing the specific texts they derived from, and in that sense they are not, strictly speaking, exegetical, for they unfold new compositions from these texts into a greater, fuller context in and for the present, as does divine history itself. Plenitude is their greater message, not scholarly precision. They provide some of the basic tools of topical argument, including – even especially – for *argumenta fidei*, which place before our mental eyes those things which cannot be materially seen but when ardently envisioned with hope and faith are witnessed and so may be confidently believed.

The Ottonian historian, Richer (who studied under Gerbert), praises how Gerbert taught the *artes liberales*, including the Topics, introduced last in the order of the *Organon*, as a prelude to Rhetoric before advancing to the capstone of all language arts, practice in actual Disputation:

> Finally [when teaching the *Organon*] he initiated his listeners to the Topica, which are the seats of arguments, translated by Cicero from Greek into Latin and explained in a commentary of six books by the consul Boethius.[8]

After Rhetoric, his curriculum proceeded directly to *mathesis* (that is, arithmetic and geometry as presented in Boethius's *De institutione arithmetica*), then music, and finally astronomy, which Gerbert taught, Richer says, with the aid of a sphere (likely an armillary) on which he had indicated arcs of the Great Circles and placed various *fistulae* ("tubes" of some sort) in order that beginning students (the *indocti*) could see, understand, and remember the difficult figurations involved in spherical geometry and astral calculations.[9]

7 See Carruthers 1998.

8 Richer, *Histories*, III.46: "Inde etiam topica id est argumentorum sedes, a Tullio de Greco in Latinum translata, et a Manlio consule sex commentariorum libris dilucidata, suis auditoribus intimavit" (my translation). On the fundamental role of disputation in medieval vernacular literary and court cultures from the time of Alcuin on, see Kempshall 2008, Novikoff 2013 and Swift 2008.

9 *Ibid.* III.51: "Cuius instrumenti ratio in tantum valuit, ut ad polum sua diametro directa, ac semicirculi productione superius versa, circulos visibus inexpertos,

Basic Shapes: Forma Quadrata and Cubus

Logic texts employed geometric forms as essential to their demonstrations, plotting the causal relations of propositions and proofs most frequently on the sides and angles of a triangle or a square. Circles and arcs were used in cosmological demonstrations and calculation, such as of terrestrial and celestial harmonies, and the nature of angels, and the several medieval treatises on spherics. The basic shapes were often combined to show highly complex logical proofs, such as the relationships of the macro- and microcosm.[10] Some early medieval manuscripts containing basic Aristotelian logic texts, best known before the thirteenth century in Boethius's translations and adaptations, do not have graphic figures drawn in them, but this certainly does not indicate that those medieval students made no use of material visualizations of their texts when learning and using their geometry (and their numbers and letters too). Textual evidence is plentiful that people used ephemeral, erasable, and thus reusable surfaces – sand, wax, pebbles, bones, and other counters such as (frequently) their fingers – as instruments to calculate, to help them when learning their "basics," and also, as we saw in the first chapter, when drafting their new verbal compositions using wooden tablets with waxed surfaces, which were designed to be easily erased. As material objects, such tools no longer recognizably or readably survive, not even Gerbert's spheres, though early armillaries and astrolabes certainly do, together with manuscript drawings that show how to set up Gerbert's abacus table.[11]

---

scientie daret, atque alta memoria reconderet." Richer says that Gerbert made at least four types of spheres for teaching, including what seem to modern historians to be an armillary sphere and a hemisphere (a type of astrolabe); see Darlington 1947, Burnett 1997, and Truitt 2012, 2015 for additional considerations of Gerbert's spheres.

10 The literature about this cosmological geometry is vast. It could – especially after the fourteenth century – take on an esoteric, even magical cast in addition to demonstrating concepts in natural physics and medicine. Places to start include Obrist 2004, Lüthy and Smets 2009, McNamee 2016, Chajes 2020, Rampling 2020. Wallis 2015 offers some well-advised cautions about reading medieval diagrams. See also the essays in Kupfer et al. 2020; and in Hamburger et al. 2022.

11 Burnett 1997 gives a fine account of Gerbert, his contemporaries, and the dissemination from Arabic texts of early medieval abacuses and astrolabes, with several early western manuscript illustrations. The Roman system of finger-reckoning widely used in the medieval world also is described in B. and R. Williams 1995; medieval reckoning also sometimes used Greek letters. Bede describes both at the start of *De temporum ratione*, 1–2 (trans. Wallis 2004, pp. 9–13). A French manuscript of Boethius's treatise made around 1100, now in the British Library (BL MS Royal 13.A.xi, f. 33v; viewable online), shows the common finger gestures used from Roman times through medieval centuries and later.

One such basic logic tool, a figure ubiquitous even in early manuscripts, is known as the Square of Opposition (see Figures 15A and B), used to indicate the possible logical relationships of a given statement, the figure's *pro-positio*.[12] It is referenced often by Aristotle as a *schema* (as in *Prior Analytics* I.4 26b) and usually translated as *figura* in Latin; in *De interpretatione* 22a, Aristotle refers to setting out a "table," *ὑπογραφεύς*, of possible logical contingencies.[13] These teaching works were incorporated into the set of elementary texts known as *Organon*, the logical "tools" which concluded with Cicero's *Topica*, and after the sixth century also Boethius's *De topicis differentiis*, taught as prerequisites to Rhetoric. *On Interpretation* was taught second in the curriculum of Dialectic, following the basic *On Categories*, which from late antiquity had included a tree diagram.[14] Third came *Prior Analytics*, and it is here that we most often find the medieval Square of Opposition figure, though it should be noted that others of these texts also described figures in order to comprehend, work with, and retain the methods taught. My interest here is chiefly in how square *figurae* were imagined (and sometimes drawn) for a variety of uses and purposes and not just for categorical syllogism, but it is well to begin with this very common and accessible example.

In Martianus Capella's *Marriage of Philology and Mercury*, the figure of Lady Geometry (whom I discussed in Chapter 3), is preceded into Jove's hall by Dialectic, the second of the Arts (after Grammar), summoned to help celebrate the grand marriage. She presents an outlandish figure, speaking strangely in words that the resplendent gods can hardly understand even in their exalted communal tongue. Dialectic's language is the opposite of common. She carries in her left hand a hissing venomous snake and in her right a set of

12 A good explanation by T. Parsons of this archtypal figure in Dialectic can be found online at https://plato.stanford.edu/entries/square/ (Summer 2017, accessed March 18, 2021). On the uses of categorical and analytical Trees, see Worm 2014, Even-Ezra 2021.

13 See Liddell & Scott, s.v. *ὑπογραφεύς*, 2, citing this passage in Aristotle.

14 The so-called "Porphyrian tree" was named from the third-century commentator on Aristotle's logic texts and author of the introductory work, *Ysagoge*, for which this diagram was created. Along with squares, branching "tree" diagrams are among the earliest and most common of reasoning tools; see Obrist 2004, Worm 2014, Even-Ezra 2021 for medieval examples and adaptations. On the common use of such teaching aids in antiquity see Netz 1999. *Schemata*, like *figurae*, are always visualized, with or without using material aids such as an abacus (or one's fingers). See the comment of Gerbert of Aurillac about learning to visualize and work with cubic shapes (*Geometria*, see Chapter 3, above) and with spheres (note 9, above).

tablets on which are sketched numerous *formulae*.[15] She resembles a cube herself, for she is square-ish of body, dark, and covered with a great many hairs like bristles and thorns (either upon her body or as a cloak – the narrator says he cannot make her out very clearly). These are perhaps the imagined *lineae constrata*, the "thatching" by which, both in practical measuring and in Gerbert's school geometry, we can calculate her appearance (her *superficies*, "surface").[16] She speaks obscurely in inexplicable words that only puzzle the wedding guests, saying things like "the universal affirmative [is] diametrically opposed to the particular negative."[17] She grew up in Egypt, thence migrated to the school of Parmenides, before fetching up in Athens with Socrates and Aristotle (¶330). The feasting gods at first laugh merrily at her appearance and her words, which are evidently just gibberish to most of them (she speaks severely, Martianus writes, "in twisted knots [=*contortis* ... *nodis*]").[18] But Athena silences the court, and Dialectic is allowed to speak, though only for a time. She is finally interrupted by the bridegroom Mercury himself, who says that the wedding party can stand no more of such difficult, dark speaking, and Dialectic is made by Athena to stand aside for the eloquent, brilliant, and far more persuasive Rhetoric, who does not bring any obvious snakes (Dialectic tries to give her vicious snake to Grammar, but even she cannot control it, and hands it off to Athena: ¶333).

While addressing the festival-goers, Dialectic tries to demonstrate her obscure sentences with the aid of a *figura*, a Square. She starts her explanation of how to reason through logical categories, modeling her words on *Prior Analytics* I.4 (26b), with what she calls a *proloquia*, a visualized "prologue" expressed as a squared shape (*quadrata forma*) made from four lines. On

15 Stahl and Johnson translate *formulae* as "patterns" to be used in analysis, which can be verbal or visual and usually both at once. Grammatically, *formula* is the diminutive of *forma*. Perhaps we could understand them simply as "models," both as sentences and as figures, like those so plentiful in the *Organon*.

16 It is worth noticing again that medieval academic geometry texts, in several ways (and without remarking on their sources) incorporated measuring practices especially from surveying and architecture into their presentations of the abstracted Euclidean figures with which we are now more familiar; see Kidson 1990, 2003.

17 Martianus Capella, *De nuptiis*, IV.327–9; trans. Stahl and Johnson, pp. 108ff.

18 Martianus Capella, *De nuptiis*, IV.327: "Haec [=Dialectica] quoque contortis stringens effamina [=speaking] nodis," a good description of the extreme compression and abstraction of the axiomatic *brevitas* of formal logic (as compared to Grammar, who has just preceded her, and to Rhetoric, who follows). Dialectic's vicious snake was replaced in some later iconography, sometimes by a barking dog's head, sometimes by a spiked mailed fist, but her graceless combativeness and abrupt, obscure language remained essential to her nature.

these four lines the possible expressions of the proposition, positive and negative, universal and particular, are placed, and then two diagonal lines are drawn joining corners 1 and 3, 2 and 4. But these numbers are given as such only in Stahl and Johnson's English translation, for Martianus couched these instructions only in verbally descriptive terms like an ekphrasis, not using any Latin numerical signs to label the corners. Dialectic speaks in terms of "superior" (*superioris*) and "inferior" (*infra*) angles, of the two diagonal lines being "led" (*ducantur*) from the universal affirmative to the particular negative and from the universal negative to the particular affirmative.[19] No wonder even the gods found this language difficult to follow, and no wonder as well that some visualizable aid was greatly to be desired (such as those in Figures 15A and B). Notice also that in Dialectic's description of her syllogistic square, the verbs used are in the present tense, "exprimatur," "ducantur," as though she were coaching someone presently trying to visualize the figure rather than describing a completed thing. And of course the instructional value of her exercise lies exactly in her audience's *acts of visualizing* it as she speaks. This is why the verbs used are in the present tense.

It is also worth pondering, I think, why the modern translators (and they were hardly alone) found it necessary to depart entirely from Martianus's Latin text, and instead to use numbers to label each of the corners in order to make the figure intelligible to modern students. It suggests that, especially when just learning our analytical skills (as elementary dialectic is now called), we require to use a fully drawn and labeled figure in order to understand the relationships discussed, rather than easily being able to imagine ourselves moving about within and over the shape, across, above, below, and transversely. Two versions of the logic Square, drawn in a manuscript of the mid-ninth century (which was written in France, perhaps at Fleury)[20] demonstrate that the figure, even when drawn *in plano*, was not conceived to be simply static. The text is

[19] Martianus Capella, *De nuptiis*, IV.400–1: "quattuor lineis quadrata forma exprimatur. in primo angulo superioris lineae scribatur universalis dedicativa [=affirmative] et in alio eiusdem angulo abdicativa [=negative]: item infra ad primum angulum particularis dedicativa, ad angulum reliquum particularis abdicativa; deinde ducantur angulares lineae ab universali dedicativa ad particularem abdicativam et ab universali abdicativa ad particularem dedicativam." I discussed the grammatical force of similar present-tense instructions for visualizing shapes in Carruthers 1998, esp. in reference to Hugh of St. Victor's *Libellus* describing Noah's Ark as a geometrically imagined instrument for meditation.

[20] University of Pennsylvania Schoenberg Library MS LJS 101, f. 37v; the colored version is on fol. 36v of this same manuscript, written at the same time. The digitized manuscript can be viewed in full online at http://hdl.library.upenn.edu/1017/d/medren/9951865503503681.

Boethius's translation and shorter commentary on *De Interpretatione*. Figure 15A shows the Square's four axiomatic types of proposition – universal and particular, affirmative and negative – together with a model sentence of each type. On the verso side of this folio, shown in Figure 15B, this same geometric figure was drawn and colored to show how the logical contingencies move transversely across the Square continuously from a superior to the diagonally inferior corner (and vice versa).

It is a small observation to add to many similar ones, but reveals again how fundamentally *dynamic* geometry was conceived to be in medieval thought, not only in creating the natural world but for human understanding and inventing too. To understand and invent properly with them, you must move about within the figures and also turn them about, retracing and "leading around" (*circumducens*) in your mind's eye the movements that create them. A much later quadrature, produced when scholastic logic reached its zenith (or nadir, as its many critics thought, including the sixteenth-century logic teachers Agricola and Ramus),[21] implies far more complex movements to be "drawn from" (*deducendi*) within it (Drawing 2). This is a *pons asinorum* figure ("bridge of donkeys"), so called (perhaps) from its service as a test to separate middling students from those advanced enough to proceed in advanced logic. Following the traditional view of logical diagrams, it is explicitly defined by the Paris logician Pierre Tartaret (d. 1522, who likely drew this version of it) as a tool of *ars inveniendi*, a visualized instrument to be studied and applied for the purpose of finding and constructing, and Tartaret does not here call it an *ars judicandi* at all.[22]

One late medieval scholar enlivened his version of the textbook Square in the margin of a copy of *Organon*, by drawing the engaging figure reproduced in Figure 16.[23] The manuscript is from the later fourteenth century, but

[21] On these criticisms, see Mack 2011.

[22] "Ut ars inveniendi … sit facilis plana atque perspectiva ad manifestationem ponitur sequens figura que communiter propter eius apparentem difficultatem pons asinorum dicitur [So that the inventing craft may be readily clear and seeable for demonstration the following figure is brought out, which commonly because of its evident difficulty is called bridge of asses]": Tartaret 1500, fol. 73 (the figure is on fol. 73v); he is commenting on *Prior Analytics*. There is mystery concerning where (and when) the term was first applied. Influential early modern school masters were highly critical of such academic complications: see esp. Mack 2011 for an overview.

[23] Montpellier: Bibliothèque universitaire historique de médicine, MS H 162, fol. 146v (a compendium of basic logic texts). This drawing is in the *Prior Analytics*; another Square, also for the *Prior Analytics*, is on fol. 152, where it appears to have been drawn at the same time as the text, not to be the work of a later hand. Digital images of these two folios can be consulted online through the Bibliothèque Visuelle des Manuscrits Médiévaux, https://bvmm.irht.cnrs.fr.

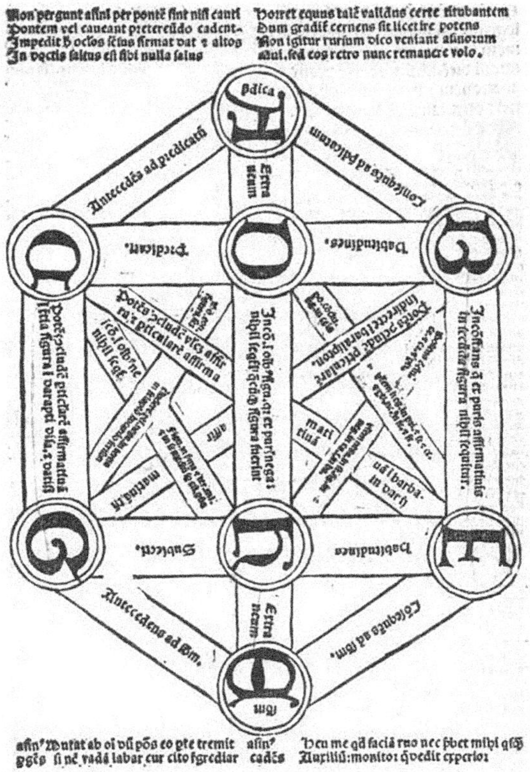

Drawing 2 "*pons asinorum*": P. Tartaret, *Expositio super textu logices Aristotelis*, fol. LXXIII verso (Lyon, 1500). Early European Books Online.

we cannot precisely date this doodle, let alone know what exactly was in its maker's mind, but filling out the logic Square as the shield of an armored knight (note its chain-mail leggings) certainly accords with the notion of Dialectic as a closed fist and logic as a form of combat. Yet this shield also

---

Many thanks to Prof. A. Even-Ezra for first showing this remarkable drawing to me. We should not assume the doodle to be the work of a bored beginner, however, for the Dominican convent at Toulouse is known to have encouraged the use of visual figures for various learning and inventing purposes: see Hamburger 2020, pp. 248–53. Dominican education practices, including using visual figures, are thoroughly discussed in Mulchahey 1998. O'Daly 2015 examines medieval manuscripts of basic rhetorical treatises, *De inventione* among them, containing many logical schemata that are integrated with the texts.

has grown a great many flowers and a long, leafy branch as a sort of plume, perhaps because from Dialectic Rhetoric produces her flowers of eloquence, or perhaps because completed ideas will blossom from the seed of arguments Topically discovered? Whatever his idea, this reader was letting his mind unfold wittily in the margin of the text he was studying, inventing from the familiar lines of the basic Square.

Such geometric figures are not found only in school texts. They can be recognized in manuscript illustrations also, just as evidently as they are in Villard's drawings, to demonstrate the creative energies implicated in school-taught geometry. Two examples will show this; both are from exegetical manuscripts of the twelfth century, each illustrating a "geometrical" in Scripture, the Ark of Noah in Genesis and the Heavenly City in the Apocalypse. Each structure is described in the Bible as a solidly fabricated edifice, with multiple dimensions, including that of time (as the sites of the First Covenant and its final Fulfillment), that preclude understanding their illustration in these pages as only pictorial linear glosses of literal text.

The Ark of Noah shown in Figure 17 is in a twelfth-century manuscript from Regensburg, one of the great monastic libraries of the time.[24] The main text in the book is Honorius Augustodunensis's *Imago mundi*, a widely known exegetical text (Honorius himself most likely was a monk in Regensburg). However the text on the page with the picture of Noah's Ark is not specific to *Imago mundi*, but is the general gloss current at the time concerning the Ark's measurements and its inhabitants, which incorporates material from previous commentaries, including that of Hrabanus Maurus.

Various clues are given to the solid character of this representation of the Ark, including the two dragon-heads at either end of the boat and the rippling seas upon which it floats. But the main clue to knowledgeable viewers is the use of double lines throughout, even in the depictions of the boat floating on the waves, but especially around the various quadrate forms that make up the modules of the Ark's three storeys and the compartments within them. The double-lined "*arca*" figure in St. Gall MS. 830 (Figure 18) is an example of a "superpendicularis figura quadrata." In this mid-eleventh-century manuscript, brought to the monastery by its abbot, Ekkehart IV, the Boethian text introducing the various double-lined *figurae solidae* explains that this one is the figure for "*recta linea super rectam lineam stans* [a straight line standing over a straight line]," the outer line being called the *superpendicular*, for it hangs

[24] This Ark (sometimes called the "Munich Ark") is now in Munich, Bayerische Staatsbibliothek MS Clm 14731, fol. 80. The microfilm has been digitized, and is viewable online.

above the inner quadrate. All the angles of this shape are right angles; in other words, it is what we commonly call a rectangular box.[25]

The Genesis text makes it very clear that the animals are *inside* the Ark, and so the artist has drawn his cubes using the framing, double-line convention that indicates that a drawing is to be understood as depicting solid forms. The storeys of the Munich Ark are indeed drawn on a flat page but are not therefore to be understood as only two-dimensional. The artist of the Munich Ark picture is invoking some commonplace geometric *figurae* to communicate to his clerical audience (who could be counted on to know this convention) how his picture should be seen and understood to provide a careful gloss describing how to visualize the structure measured out in the Genesis text.

It is of a piece, I think, with the common medieval habit of describing diagrams, which we are now taught to comprehend only as finished and exclusively visual objects, in terms of instructions to students to mentally draw and move about within them *right now*, their temporality indicated by the tenses used in the texts, past tenses for looking at a completed object or action, but present and future (including imperative verbs) for actively visualizing, and thus *ante oculis ponendo*. The drawing itself presents both *argumentum* and *instrumentum fidei*. Once again the logic of verbal deponency, the *passiones* (as grammatical inflections) commonly used to command and express the activities of imagining, serves to encapsulate the nature of the mental activities involved, the mind actively picturing with its mental eyes a new composition made from images and words recalled inventively from its copiously inventoried stores.

Gerbert of Aurillac uses a different planar figure than did the Munich Ark painter to indicate a *figura solida quadrata*, one known now as an embedded square (Drawing 3).[26] It consists of two squares, one drawn parallel to and inside the other. Four diagonal lines are drawn from the corners of the outer quadrate to the corresponding four corners of the inner one; in some manuscripts the figure is just a single square with a diagonal drawn from each corner toward the center, but not crossing it.[27] Figure 19 shows a picture of the Heavenly City

25 I have quoted the text from lines 15–18 of page 295 of Cod. Sang. 830 (Figure 18): "superstat linea super eam quam insistit [lies within] sup*er*pendicularis uocat*ur*." Such figures are not only squares, as the text also makes clear, for any double-lined figure is understood to be three-dimensional, a *figura solida*, not *in plano* (flattened).

26 On how a basic embedded square was commonly laid out by Gothic builders in order to true the measures of their surveyed plots, see Murray 2014, esp. pp. 37–46, 100–2. Many thanks also to Megan McNamee for several conversations about the use of Gerbertian geometric figures in the two paintings discussed here, and to Adam Cohen for introducing me to the Munich Noah's Ark.

27 McNamee 2016 discusses this figure as indicating a solid form.

Drawing 3 An embedded square. From a manuscript of Gerbert of Aurillac's *Geometria*, written in Bavaria, mid-twelfth century. Philadelphia, University of Pennsylvania Schoenberg Library, MS LJS 194, fol. 4v (detail). CC-BY-4.0

described in John's Apocalypse, from the "Beatus of St-Sever," a manuscript, now in the Bibliothèque nationale de France (MS Lat. 8878), that was made in southern France, just over the Pyrenées from Léon, at an abbey in Gascony in the later eleventh century. This painting was made as a meditation device to accompany the best-known commentary on the Apocalypse at the time, written by a ninth-century monk in northern Spain named Beatus of Liebana. When I first considered this picture I could not understand why there were heart-shapes at each corner. There seemed so many allegorical possibilities, but none very likely, especially since the shape we now conventionally call a "heart" was not called that until the fifteenth century.[28] But when I showed it to practicing artists, they immediately recognized the figure as representing a cube. "Oh," they told me, "those shapes indicate an action of folding together, that the walls should be understood to be rising up around the inner square." This of course had never occurred to me, being but a historian of words. And I still did not really understand it until I came across the embedded square figure in Gerbert's text, and recognized in it the underlying geometry of this painting.

The outer square shape is the standard planar (flattened) figure of a cube, in the five quadratures, according to Roman surveying method, which together

[28] The history of this shape is discussed by Jager 2000. My thanks to Anne and Patrick Poirier, who were resident artists at the Getty Institute in the same year I was, and from whom I learned a great deal about figuring geometry as well as memory.

compose the surface area of a cube or *cella*, when rendered in a planar perspective. The heart-shapes represent areas of overlap when the flattened cube is folded into its solid form, and signal to viewers how the painting is to be properly understood. For when these corner "hearts" are imagined as folded together along their prominently drawn diagonals, the walls will indeed rise up on the four sides of the inner square to form a cube. Notice how the diagonal is emphasized in the painting by the angel's measuring-rod. If you should try to construct a box from a flat piece of paper, you will likely proceed more or less as this figure requires that you do, by drawing a central square or rectangle, then making a fold or perhaps a cut between each of the inner corners and the corresponding outer corners of the paper. As the four sides are folded up along the diagonals, they will also fold against their adjacent sides to form the four corners of a box.

The same basic geometry is at work in a Heavenly City painting designed by the monk Maius in the mid-tenth century, a decade or two *before* Gerbert composed his geometry treatise, in a manuscript which is now in the Morgan Library (MS M644).[29] Since the embedded square figure was introduced to northern Europe from Arabic geometry texts, transmitted via Egypt, north Africa, Spain, and Catalonia, its use as a suitable figure for a solid building in these paintings, made in monasteries mostly in the region of Castille y Léon even before Gerbert wrote, makes sense.[30] An embedded square is the figural basis also for depicting the Heavenly City in the twelfth-century Beatus made at the Benedictine monastery of Santo Domingo de Silos, near Burgos (Figure 20). Notice again, how in these various paintings the diagonals are greatly emphasized by the angel, both in directing John's gaze and by how he holds his measuring rod. One might suppose (correctly, I think) that those angelic diagonals offer us our starting principle (*principium*) for comprehending the figure. In these paintings, the cube's top is our viewing station, from which we

[29] Shailor and Williams 1991. On the Latin translation and dissemination of Arabic mathematical texts, see Burnett 1997, 2009.

[30] Shailor and Williams 1991; also Sanchez Ameijeiras 2019 on sculptured movement. On Maius, see esp. J. Williams 2017; he argues persuasively that Maius is likely to have worked at Tábara in southern Léon, and to have been influenced by Gascon styles in monastic communities around Tours as well Mozarabic ones. Morgan M644 is discussed in detail, pp. 69–72, Williams commenting that "Maius seems to reveal some sense of the plasticity that the rest of the school ignored," for his figures seem "to inhabit comfortably an illusionistic space, had the style recognized such a concept" (p. 72). But Maius likely had learned a better way to represent a *quadrata solida* from Arabic-Spanish sources. S. Murray 2014 discusses the use by masons of square figures (resolvable into triangles) to lay out the basic shapes of their buildings, pp. 37–46, 100–1 esp.

look down upon the crystalline city whose walls are as clear as glass (Apoc. 21: 16–18 esp.). This is also the conventional viewing point of katascopic philosophical meditations, the Geographies, Topographies, World-Maps that are such common geometries for ancient philosophical speculation.

In twelfth-century Paris (for this most useful figure soon made its way generally), Hugh of St. Victor describes a variation of this elementary geometry in his "little book" for constructing meditative thinking, by carefully imagining a complicated and comprehensive figuration of the Ark of Noah (*Libellus de formatione arche*). This figure has been drawn graphically for students now, using tools such as computer-assisted drawing (CAD) and modern printing techniques. The details of how it might be imagined are now readily available, and essential for modern readers untrained in visualizing solid forms only from planar descriptions and mentally seeing them even in differing perspectives. Hugh's careful instructions pay great attention to the diagonals in the figure. When its basic planar features are complete, Hugh instructs students to then imagine the central column of the Ark as raised up ("cubitus ille … sursum elevetur"), understanding that as they do so the entire structure will rise as though pulled up from above (*sursum*) to form a solid figure of the three-storeyed Ark. One then should look and move about within the Ark, climbing up the steps (*gradus*) of its different ladders (*scalae*) as well as examining its various chambers. The exercise stresses moving from location to location, in time as well as space, as suits a meditation focused as much on salvation history as this one is. An alternate name for the treatise is *De pictura arche*, and it is indeed a rhetorical *pictura* to be imagined in all its detail, *pictura* being at this time a synonym for the ancient literary trope of ekphrasis.[31] The basic geometry it describes is an embedded square (or, rather, a series of them, each inside the other).

The cube (*cella*) in monastic contemplative writing was a particularly productive three-dimensional human space, and not, I think, only because it was the basic module of Romanesque monastic architecture. Peter of Celle, the Benedictine abbot of the late twelfth century whom I briefly introduced in my first chapter, was a friend of John of Salisbury and of the great Victorine

[31] See the detailed discussions in Carruthers 1998, 2008; an English translation is Carruthers and Ziokowski 2002, pp. 41–70, with a selected bibliography. Rudolph 2014 gives a number of possibilities, drawn *in plano* from different angles and in varying detail (Rudolph rightly says, p. 3, that although his graphs are of necessity published *in plano*, they are meant to be understood as solid – Hugh himself makes a similar comment in *Libellus*, ch. 1). Though modern scholars have attempted other graphic renderings of Hugh's *pictura*, none of the many extant medieval manuscripts of the work does so – this fact argues that it was understood as an ekphrastic "picture," not as instructions for a particular material work.

contemplatives. He makes clear that it is the *shape* and measure of his cell that enables his meditating mind:

> After my daily burdens, and thirsting in some measure for the silence of my cell, I draw deeply from the quiet which has now been granted me. The mind has a more extensive and expansive leisure within the six surfaces of a cell than it could gain on the outside by traversing the four parts of the world. In fact, the smaller the place the more extended the mind, for when the body is constrained the mind takes flight. A limit upon the body expands the mind. If no one interrupts, my mind leaps as high and as far and as deep as it wants.[32]

But the *cella* Peter counsels his brothers to use for themselves is in their imaginations, for like most Benedictine monasteries of the time, the brothers shared a common dormitory and refectory. Each must learn how to withdraw into a quiet, solitary place in his mind, imagined in the productive form of a six-sided cube (for his meditation *cella*, like any monastic room, has six surfaces, including a ceiling); each can (and should) imagine walking about within it, and so viewing the topical matters and their arguments for his thinking from several changing perspectives, as he may assign them. The meditation cell seems to be simple, imageless, even barren, but when used intelligently it can become a contemplation machine, hoisting one's thought to greater and higher matters. It remained a favored shape for meditation in the centuries to come, as it had been for centuries before.

Basic Shapes: Rotundities and Wheels

The sphere is the primary form of the cosmos, the shape that contains all circles and linear solids, as Martianus says (as do *Timaeus* and Calcidius).[33]

32 Peter of Celle, *De afflictione et lectione*, ed. Leclercq 237–8, trans. Feiss 139: "Grata est quies post laborem. Crudeli certe et continua uexatione toto anno oppressus, indulta nunc quiete, uel ad modicum cellae silentium quasi portum desiderabilem multum sitiens haurio et largius atque latius otiatur animus infra sexquipedales forte parietes quam extra discursando quacumque per quatuor mundi partes. Proinde quo angustior locus, eo diffusior animus, quia quando premitur et refrenatio corpus eleuatur animus carnis dilatatio est mentis. Exilit enim, nemine prohibente, quarn longe, quam late, quam profunde uoluerit."

33 Martianus Capella, *De nuptiis*, VI.722: "sphaera sane intrinsecus capax omnium circulis subsistit, in quos resolvitur, soliditas vero efficit schemata generalia, ... item prisma, id est sectio [=a cut], quae instar [=image] schematis est; item cybos, item conus, item cylindrus, item sphaera ... quae cuncta ut ordine suo monstremus in pulvere." Notice how Martianus stipulates that these figures, visualized in cross-section, are what we demonstrate [*monstremus*] by drawing them in sand

It is the animating movement from which all further expression, physical and human, derives. I have discussed its use in meditation at some length already – in Chapter 1 as the organizing shape for Nicholas of Cusa's "On a Bowling Game" (*De ludo globi*), his ethical treatise on virtuous governing both of society and self; and in Chapter 3 as an instrument for cosmic meditation in Alan of Lille's unfinished work "On the Intelligible Sphere."

These examples demonstrate well the powerful inventional capability of diagrams when used as tools of calculation and computing. I do not mean computation only in the sense of numerical calculation per se, the sort of calculating for which an abacus or an astrolabe is so very well designed, but more inclusively as the matrix of logic, in fact of topical arguing and measuring of all kinds. The motion implied in one geometric figure was considered to be particularly well adapted for this kind of syntactic work, namely, the line curving round in a circle, in its solid forms *sphera* and *globus*. Exegetically, as we saw from the complaints of the Liège geometers, such created shapes were notoriously (at least according to Franco) considered to be *imperfect* "rotundities"; their trait of "roundness" (*rotunditas*) records their essentially *geo*-metric condition. Just as a standard modern soccer ball is not a perfect sphere, so all geometric "rotundities" are constructed through linear forms: that is the mark of their being earthly globes.[34]

In a telling counter-example of this unmathematical conception of "roundness," Roger Bacon (in the later thirteenth century), commenting on the Exodus account of the furnishings of the Temple (the Menorah in this case, Exod. 37: 17–22), describes in some detail how properly to visualize the *spherulae* of each branch of the seven-branched *candelabrum*, taking issue sharply with what was then a traditional gloss:

> For because of the ignorance among theologians of [Euclid, Theodosius of Bithynia, and Menelaus of Alexandria, all of whom wrote on the sphere], Scriptural texts are misunderstood not only in greater ways but in lesser

---

[*in pulvere*], just as Geometria does here. See also Figure 21, showing Geometria with a round board, though an abacus-board was often four-sided. Much has been written about this ancient idea; a recent account of Roman versions in Ovid and Cicero is P. Kelly 2021.

34 An illuminating (and highly readable) account of soccer balls and other such polyhedronal *sphericulae* as topologies is Kotschick 2006, available online at https://www.americanscientist.org/article/the-topology-and-combinatorics-of-soccer-balls. A standard soccer ball is stitched or glued together from 32 polygons, 12 of them five-sided and 20 six-sided, arranged in such a way that every pentagon is surrounded by hexagons. Linear forms, being composed of solid elements are all made up of triangles (measuring their three dimensions): see Plato, *Timaeus* 53d.

> ones. For those with learning only from the histories say that the *spherulae* of the candelabri were round bodies but not having complete sphericity. But they are not called lesser spheres for this reason, but because they were small spheres having the complete definition of a sphere as defined in Euclid and Theodosius. For the Jews in their language called these "*spherulae pomeas*" ("apple-like") from their apple-like roundness and sphericity, and likewise with infinite others. O how ineffably the beauty of divine wisdom might shine and its infinite benefit abound, if these geometricalia which are contained in Scripture, were placed in their corporeal figurations before our eyes.[35]

What this Baconian quibble with those interpreters he considers unlearned suggests to me (recall Franco's rhetorical quarrel with those ignorant theologians who despise geometry for being somehow an impious study) is that in common speech geometric "rotundities," like "quadrates" and "cubes," admitted a range of variations that our exacting mensural expectations can no longer tolerate.[36] The theologians were being cautious, but not primarily from ignorance of Euclid: rather, because, though God commanded that the Exodus shapes be made, it was Bezelel, a human craftsman and his human atelier,

35 Roger Bacon, *Opus maius pars* iv.5, "Mathematicae in divinis utilitas," p. 210: "Nam propter horum ignorantiam apud theologos non solum decipiuntur in maximis, sed in minimis. Dicunt enim cum magistro in historiis quod sphaerulae candelabri fuerunt corpora circularia, non habentes plenam sphaericitatem. Sed non dicuntur sphaerulae diminutive propter hoc, sed quia fuerunt parvae sphaerae, habentes tamen completam definitionem sphaerae Euclidis et Theodosii [=Theodosius of Bithynia]. Nam Hebraci in Gallico suo [=in their vernacular] illas sphaerulas vocant pomeas a rotunditate pomali et sphaerica, et sic de aliis infinitis. O quam ineffabilis luceret pulchritudo sapientiae divinae et abundaret utilitas infinita, si haec geometricalia, quae continentur in scriptura, figurationibus corporalibus ante nostros oculos ponerentur." (My translation.)

36 Nicolas of Cusa expounds this Boethian idea in his "Game of Spheres" when his persona in the dialogue, *Cardinalis*, explains that "the roundness of one round actual thing is rounder than another" ("Game," I.16). In Neoplatonist terms, round objects, including our mental envisionings, all only "participate" in the absolute Idea of sphericity, of which they are all only "images." Bacon shows a certain rationalist and scholastic confidence here in human abilities to penetrate, at least to a degree, even the truths of divine *Arithmos*. On Bacon's practice of what he called *scientia experimentalis* and his non-university, even papal sponsors for this work see Truitt 2021. Tachau 1998 discusses the suspicions and rivalries apparent in Paris at this time between theological historians and geometrical rationalists as these are reflected in the lavishly illuminated thirteenth-century *Bible moralisée* manuscripts commissioned for the royal court. The political dimension of these academic suspicions as they involved rivalries of the French court and the Vatican should not be overlooked.

who actually made them. And while God's Ideas articulated in his scriptural commands are always perfect, earthly efforts to realize them cannot be. The deliberate definitional caution of the historian-theologians also supports, I think, Madeleine Caviness's explanation of the so-called amateurish and inexact drawings of familiar geometrical shapes in the *Scivias* envisionings, as being Hildegard's and her nuns' human efforts to represent the perfect creation of divinity through geometric figures that to us necessarily seem distorted.[37]

Geometry, we must remember always, derives its very name from the Earth and its perspectives. As such, its methods could never completely measure Heaven. Recall how, as the pilgrim-Dante approaches the very limit of his *alta fantasia*, he likens his effort to visualize and understand the mysterious figure of the Trinity that he sees to the mental straining of a geometer trying to measure the area of a circle ("misurar lo cerchio," *Par*. 33. 133–4). In Archimedean geometry this is done by multiplying the straight lines of a polygon within the circle's circumference until it seems to be – but never quite is – coincident with the circle's area. A human geometer can only calculate the area truly by using the value of *pi*, a value that stretches incalculably and thus *praeter rationem* ("beyond reason") into Eternity. A human geometer can only see the entirety of the divine sphere with what David Albertson has called "a gift of ecstasy" – a *raptus*, in Nicolas of Cusa's term, like the *raptus* that "seizes" the mind beyond its rational powers into God.[38]

A common iconography of *Geometria*, as she is depicted in British Library MS Burney 275 (Figure 21), shows her holding her *abacus*, the sand-table by means of which geometric figures were commonly demonstrated. Martianus Capella describes her as using one the size of a large table.[39] In the Burney image, her abacus is circular, with a number of linear shapes depicted on its surface. It is both a common tool (a sand-table) and simultaneously illustrates graphically the maxim (also found in Martianus) that a sphere contains all other shapes within it. It implicates and in-folds them (for us to rationally

37 Caviness 1998, and on the point more generally Caviness 1983.

38 Albertson 2014, pp. 249–52, at p. 246. On the puzzle mode (a concept borrowed from Greek into medieval Latin as *aenigma*) in Greek (and especially Alexandrian) mathematics, which he rightly calls both aesthetic and ludic, see Netz 2009a and 2009b. It is of a piece with Aristotle's maxim that philosophizing begins in wondering. See also Netz and Noel 2007, esp. 27–64, 87–116. On the particular Pauline resonance of *aenigma*, see I Cor. 13: 12. I discussed Dante's particular use of "squaring the circle" in my 2021 Dartmouth College Zantop Lecture on "Dante's Intent Geometer," available on YouTube.

39 Discussed in Chapter 3, above. In Martianus's description, Geometry's abacus is rectangular, covered in greenish powder, and large enough to need carrying in by several assistants.

explicate them from it) as Geometria does in the Burney illustration. The idea that creativity is a process of explicating (un-folding) forms that are implicit in some matter (*res*) is analogous to the compositional synergy of compression (*brevitas*) and dilation (*augmentatio*, *dilatatio*) in rhetorical invention, the creative pulsing that is the fundamental energy, the *dunamis* and *anima* of our human work, shaping it into various configurations of a whole.

The *mappa mundi* ("world-map") is another common geometrical tool of invention, associated with Genesis but deriving as well from a familiar contemplative genre of ancient philosophy, the geographical meditation of *kataskopos*, a mental flight looking down from on high upon the features of the round globe, notably (for Latin speakers) in the "Dream of Scipio" myth well known from Cicero (his *De re publica*) and Macrobius's extensive re-envisioning of it, and indeed its use by Dante and Chaucer. In Christian use, as *mappa mundi*, it is not only a spatial map of physical features, but equally, even primarily, a mapping of Time, salvation History unfolding in measured Ages from Creation and Fall in Eden, then on to the Exodus, to Jerusalem, to Rome and beyond into cities and regions of northern medieval Europe. Time and Space are coequal within its circuit, while Divinity (as Christ), located within the encircling figure of the mandorla, gazes upon the world and blesses all of temporo-spatial Creation.[40]

But the round *mappa* is not the only structure ever used for such imagined geographical overviews. One exception is the *Topographia* of Cosmas or Kosmas Indicopleustes ("Kosmas-India-traveller"), the pen-name of a sixth-century syriac Eastern Christian writing in Alexandria, who describes his own earlier travels as a merchant (perhaps as a notary-agent for a merchant guild) as far east as western India and Sri Lanka, as well to various places in eastern Africa, Arabia, and the eastern Mediterranean.[41] In his Topography, he pictures the Earth's basic shape as consisting of two rectangles, one atop the other (the earth and the sky), to form a box or Ark structure, the most common exegetical figure for the universal Church founded on the historical Earth. This Ark is surrounded by water, and a great mountain rises on its north which seems nearly to touch the sky, over which the sun's daily course apparently

40 A great deal has been written about this basic characteristic of medieval maps, most recently by Kupfer 2016 and 2017, and see also Jacob 1992.

41 A modern edition by W. Wolska of the Greek text, with French translation and commentary, was published in three volumes in the SC series (Paris: Cerf, 1968–73). The text printed in the PG is from the early eighteenth century. Notice that the geometric figure Kosmas describes for his *arca* is a type of embedded square. On Kosmas as a merchant, see McKrindle's "Introduction" in Cosmas Indicopleustes, *The Christian Topography of Cosmas.*

rises and sets. As his authorities for this strange geography, Kosmas invokes Scripture, in particular the measures of the divinely ordered Ark-Tabernacle as given in Exodus 25–26. Though some have taken his work as evidence that "Kosmas" thought that the Earth was really flat, his opinion about this as physical fact is both unknowable and not material to our understanding of the Ark as a figure of the Church in history, which is especially evident in the first five books of the work.[42] The philosophical genre of *geographia* at this time was essentially contemplative and visionary, and the Tabernacle-Ark's figurations as measured out in Exodus were common scriptural *geometricalia* used, as we have seen, to structure many contemplations on religious life. It is particularly suited to the worldwide, multilingual mission of the Church to all nations, as commissioned at the Pentecost. The Churches of the East (also called "Syrian" or "Nestorian") had taken as their particular mission the territories spreading from Palestine and Egypt to the eastern coasts of Ethiopia, the Red Sea, and Yemen (the biblical kingdom of Sheba), and east to Parthia, western India, and Sri Lanka, the very areas emphasized by Kosmas. But Kosmas wrote for his fellow Christians, neither for navigators nor for ancient philosophers. His *Christian Topography* seems to me to depict primarily the historical unfolding of Earth's salvific history into the East under the figure of the Ark (whose name, whether in the Hebrew, Greek, or Latin Bibles, means "a box")[43] as the Christian mission spread eastward to save the whole world,

42 Kosmas is an elusive figure, and there is controversy about nearly every aspect of his life, especially whether he himself made the journeys to India that he claims. That some person wrote in Alexandria in the sixth century who was a clerically educated Christian scholar of an Eastern Church, "syriac" sect, seems settled, but little else, including how much of what is now called "his" book is actually by him (the vague English use of "they" as both singular and plural in reference is especially apt for Kosmas). The wide range of opinions (theological, historical, textual, and anthropological-geographic) regarding the author(s), the books' composition, and the cosmology-geography of the *Topographia* is apparent throughout its scholarly history: see McKrindle's translation 2010, Wolska-Conus 1962 (and her 1968–73 edition of the text for SC), Faller 2011, and Kominko 2013. For a consideration of the different conception, shapes, and functions of maps in late antiquity, see also Jacob 1992, Dan 2017. I discussed the scheme of the Tabernacle-Ark as an antique Christian meditation structure in Carruthers 1998, esp. 231–7.

43 As shown in Figure 18, reproducing several planar drawings in St. Gall MS 830, one of which is a framed quadrate clearly labelled "arca." The central rectangle in Kosmas's drawing primarily shows the travels and conquests of Church History, as is true of medieval *mappae mundi* generally. It is well to keep in mind that Christian exegesis conflated the Genesis-measured Ark of Noah with the Exodus-measured, gold-covered Ark of the Covenant in the figural unfolding of time, whereas Jewish exegesis did not (and indeed the Hebrew text uses a different noun for each, as the Vulgate does not).

just as the first divinely measured Ark saved Earth's creatures.[44] It is neither a cartographic spatial rendering (a *topographia* proper) nor a *geographia* in the Greco-Roman philosophical tradition, as its contemporary audiences would certainly have recognized. Indeed Kosmas, as I understand the work, parodies the pagan tradition in order to underscore and mock its errors (as he does especially in the first books) in terms of the greater, fuller vision of his own Christian faith and understanding of the unfolding of the Christian history implicit in the Ark.[45] It is also important to recognize that the Churches of the East were by then a beleaguered group, considered to be fringe outliers from Constantinian imperial Orthodoxy: predictably Kosmas's imagined topography was not taken up by the makers of western and Byzantine Christian *mappae mundi*.

There is another a favored geometric figure used to express the perfection of divine creation, the "Almond" or lens (*lenticula*) known as the Mandorla, in which, in one iconographic tradition, the Creator sits and from which he creates.[46] The shape is also that of a burning glass or, when suitably coated, a burning mirror (*speculum*) as it was also called, the convex magnifying lens

44 In his first exegetical books (I–V, considered likely to be the earlier compositions), Kosmas says that the *topographia* describes the Christian future of the world under the manifold figure of the Ark (*arca*), rejecting the philosophers' *geographia* which describes the unredeemed earth only as it is shaped presently. The Biblical Ark is a divinely ordained, life-preserving "box, chest" (*kibotos* in the Greek Septuagint of Gen. 6: 14) and a frequently invoked Pauline image of salvific Christian history (see, for example, II Cor. 5: 1, Hebrews 9). The Septuagint text also stipulates that Noah's *kibotos* is *tetragonon*, "having four sides" – in other words a box or chest, which seen kataskopically "from above" would look like a rectangle floating on surrounding waters. Of the various items within the Exodus Tabernacle, Kosmas links the Earth particularly to the tabernacle Table, because of its molded edge (like a frame), which he likens to Ocean. Book V, where much of this detail is found, provides lengthy exegesis of the Pauline doctrine that scriptural *geometricalia* spiritually measure the recurring patterns both in dimension and time evident since the divinely ordained simultaneous creation of the cosmos and of history.

45 It is tempting to name the second-century *Geographia* of Ptolemy as a particular *auctor* for Kosmas's Christian parody, especially since Ptolemy also wrote in Alexandria, also presents his sources as including merchants' travels, and extensively includes India and Sri Lanka. But there is no hard evidence for such a link. What seems certain is that the modern canard that Kosmas depicts the geographical world as flat is wrong. Kosmas's cartograph does not intend to represent the spatial world, but salvation History under the traditional Christian figure of the *Arca*, Ark.

46 The geometric figure was named from its shape, Latin *lens, lentis* being the word for a lentil seed. Medically, the word named the lens-body of the eye. Purity and transparency are its main qualities, completely captured in the Creator-Redeemer's gaze upon the world through the mandorla/lens of Heaven.

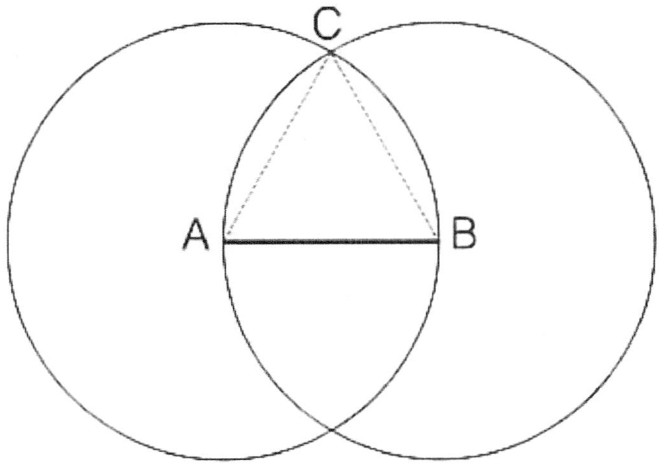

Drawing 4A Standard figure illustrating Euclid's *Geometria* I. 1

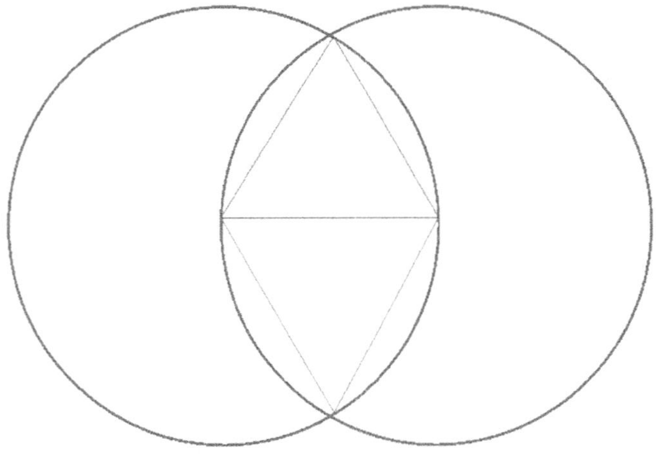

Drawing 4B Extends 4A into a *lenticula* (lens). Drawings courtesy of J. N. Crossley.

(or concave mirror) that can focus light into a single fire-producing point, analogous to the *fiat lux* of Creation itself.[47] The figure is often drawn in manuscripts of Boethian geometries (see Drawing 4A). It consists of two compass-drawn circles overlapping in such a way that the circumference of each on its circular route passes through the originating, central point of the other. The resulting geometric lens traditionally framed the "Majestas Domini," the figure of Christ in Majesty. In geometry, within the area of overlap (the "lens," as in Drawings 4A and 4B) one can then measure within the circles' arcs an equilateral triangle, as Boethius and Euclid and many others demonstrated. One can also draw, within the lens-shaped overlap, a reversed equilateral triangle of the same measure (forming a rhombus or *lenticula*), and indeed a great Many can be multiplied from that initial One.[48]

Basic Shapes: Lines and Trees, Ladders and Towers

Steps (*gradus*) and ladders (*scalae*), along with trees, were another sort of basic structure for meditative thinking. And, like trees and properly persuasive arguments, they must be climbed, one step at a time. They have been primary *figurae* used for works of and about contemplation in many religions in addition to Judeo-Christianity, both Greek and Asian. What is not always appreciated as much as it should be is that all such linear forms also require measured linear movement on the part of those using them, for they present step-by-step vertically articulated itineraries that require journeying from beginning to end, just as much as do horizontally presented itineraries, such as maps. Even the angels in Jacob's visionary dream climb up and down the heavenly ladder – they do not just float, feathered and winged though they are. Climbing a ladder requires a determined will and ardent desire to reach the goal prior to any knowledge gained. As Richard of St. Victor notes: "Nemo

[47] Just when the fire-kindling property of some lenses and mirrors was known in antiquity, and how these instruments were understood and used remains a subject of dispute among archaeologists and historians, but it was certainly remarked by Pliny (*Natural History* Bks. 36–7, discussing crystals) and by Seneca, among others. On ancient and medieval knowledge of such lenses, see Knorr 1983 and White et al. 2021.

[48] This ancient *lenticula* figure (in Drawing 4B) was demonstrated in Archimedes's treatise on geometry and is sometimes said to be Pythagorean (like so much else). It includes the figure (drawn minus the lower triangle) for the First Proposition of Euclid's *Geometry*, so this proof could be even considered as the initiation of his method, since all other linear forms were analyzed as proceeding from the central triangle enclosed within it (as Grosseteste says, *Hexaemeron* IX.1). It is now often also called *vesica piscis* (the swim-bladder of a fish), but that term was used for it only from the sixteenth century (it is associated with Dürer).

enim repente fit summus, nec scalae summitas volando, sed gradatim ascendendo scanditur. [No-one suddenly achieves the top, nor is the summit of the ladder climbed by flying, but by ascending step by step.]"[49]

The *scala*-figure structures some of the most fruitful treatises on meditative method as well as many meditations and interpretations themselves. The great Greek contemplative John Climacus gets his name from using this form for his guide to Greek monastic life, *Climax* (ca. 600; *climax* is Greek for "ladder"), translated into Latin as *Scala paradisi.* Using a mental ladder (with twelve steps) was forcefully commended in the seventh chapter of Benedict's *Rule* in order to learn the essential monastic virtue of humility. Notably, a ladder model was used in the twelfth century to discuss a mystical "ascent" through prayer by the Carthusian Guigo II, who composed his own version of *Scala paradisi.* In each case, the ladder is not only a reminder of associated virtues (and vices) but is intended also to be used as an ethical guide, directing a monk's progress to knowing God.

There is nothing uniquely contemplative about the *scala* form. It was so widely used from antiquity on because it is so adaptable; it could be invoked for any sort of movement occurring in measured stages (for instance, a curriculum in which students progress through "grades"). Both Hugh and Richard of St. Victor characterize the three Solomonic books of the Bible as a kind of *scala* (*quasi quadam* as Richard says), a ladder made from books increasingly difficult to understand that leads up to the stage of active contemplating itself.[50] Richard's works reflect this in his readers' ordering of the sequence "Benjamin minor" (*De duodecem patriarchi*) and "Benjamin major" (*De contemplatione maior*). The other title given to "Benjamin major" (*De archa mystica*) stresses the main structuring *figura* (the Ark of Covenant) used in that treatise. It seems to me that this variety of titles speaks not to scribal or librarians' confusions but to the varieties of audiences that read and made use of these works over the centuries. The titles hold clues to demographic changes in the audiences for these works, from exclusively clerical and thus Latin, to vernacular-reading laity. There is no single title for either of them, a sign of their popularity and lasting influence.

One can see these changes in audiences demonstrated clearly in the works of Bonaventure. Mark Chinca reminds us that the "vocabulary of steps and

49 *In Canticum Canticorum explicatio*, 21; PL 196.476 (my translation).

50 The term "Solomonic" for the sequence Proverbs, Ecclesiastes, and Song of Solomon is traditional (and used by Jerome); all were considered to have been in one way or another attributable to Solomon. See Hugh of St. Victor, *Didascalicon* IV.8 (Taylor notes that Jerome and Isidore were his sources, p. 217 n26). Richard's use of this ladder image is in his prologue to *In Canticum Canticorum explicatio* (PL 196.408).

ladders is conspicuous in Bonaventure's spiritual exercises, but with an altogether different meaning and valency [than in monastic orthopraxis]. To climb the steps of the ladder now means to execute the stages of a procedure which has been described in writing before it is put into practice, and is not tied to any particular form of life in any particular community, but may be applied anywhere."[51] The Franciscan order served a transitional social function during the thirteenth century, for Franciscan practices bridge, transmit, and also translate (in the manner of *translatio studii*) those of the enclosure-centered Victorines (who, true to their monastic calling to live in the extra-mural "desert" apart from "the world" of the city, refused a Paris bishop's request to take on urban parish care). Broadly directed pastoral care, including preaching, for a vernacular-speaking laity was evidently needed by the late twelfth and early thirteenth centuries as the Church spread, centralized, and laicized its authority in the world. There were many consequences of this urbanizing, including on graphic illustration. Diagrams were increasingly drawn and painted, no longer only to be imagined by trained adepts, but graphically portrayed on walls and in manuscripts to a far greater degree than they had been in previous centuries. Two works by Bonaventure give examples of this contrast, his *Lignum vitae* ("Tree of Life") and his *Itinerarium mentis in Deum* ("Journey of the Mind into God"), both employing common linear structures but intended (at least initially) for quite different kinds of audiences.

Both treatises are composed in such a way that they can be used for a variety of purposes. As I showed in Chapter 2, Bonaventure ends his account of his commemorative trip to the site of Francis's direct seraphic vision on Mount LaVerna by emphasizing that no *raptus* was granted to him, but that he is going compose a treatise for his fellows, using his own visualization of the seraph's six wings in such a way that they in turn can use the figure to support their own meditations.[52] And since, as we know from Geoffrey of Vinsauf and many before him, that Envisioning is the first necessary step in any act

[51] Chinca 2020, p. 18. Chinca begins his analysis with the spiritual writings of Bonaventure, proceeding thence to the increasingly middle-class, lay, and vernacular audiences addressed by later works, such as the Dominican prayer-book *le Somme le Roy*. Contrasting Bonaventure with Anselm's early twelfth century *Prayers and Meditations*, Chinca comments: "Whereas Anselm put the reader in charge of the text, allowing wide scope for individual adaptation and variation of the script, Bonaventure put the text in charge of the reader, whose meditations must follow the exact prescribed [and often numbered] order." (p. 57; brackets in original).

[52] I discussed the seraph figure in detail in Carruthers 2009. Geoffrey of Vinsauf's advice to all seeking to set out to invent a new work to first have a plan of it firmly laid out in one's mind as map-makers and architects do, was essential advice for artists of whatever sort: see *Poetria nova* I.55–58, and my Introduction, above.

of composing, whether for prayer, or for "tracting" in words, for "seeing" the articulated sounds of music, or for planning how to make a building or lay out a manuscript page or a painting, we should not be surprised that Bonaventure's Tree and his Itinerary both served as structural models in a variety of genres. There is no exclusive audience or purpose for either work, so neither can be classified adequately as a school diagram or a prayer diagram or a subject diagram or even an *ars predicandi*, though on occasion these two shapes (tree and journey) have served all these purposes. This is an excellent demonstration of the primary importance of its shape to a diagram made for invention and composing purposes. Unlike medical or political diagrams, invention diagrams are not abbreviated instruction manuals for studying one particular subject. The six-winged seraph figure was used from the start within the religious orders for a variety of compositions, as a most delightful and fruitful form for topical arguing and composing, since it so well satisfied the logical, rational requirements discussed in the *Topica* tradition as simultaneously *ars iudicandi* and *ars inveniendi*.

Bonaventure begins *Lignum vitae* by stating that his goal is to "enkindle feeling, shape understanding and imprint" the memories of his brethren with experiences so compelling that each may feel himself to be nailed to the Cross with Christ. And because imagination helps understanding he has bound his own passionate desires to feel and know Jesus's Passion moment by moment in simple, familiar words that are easily remembered, and has arranged the chief ones (the Topics) in the form of an imaginary tree,[53] whose twelve branches he has arranged in three groups of four topical locations each, which he calls *imaginaria*, discrete "imaginings," serving like chapters of the whole. The branches within the main three groups are arranged four to each side of the trunk, the four branches on each side labeled with individual *versiculi*, brief verses alphabetically arranged. From each branch also hangs a single fruit having a unique flavor (sweetness, humility, bitterness, etc.). After his detailed verbal picturing of this imagined tree, Bonaventure then invites his reader: "Describe igitur in spiritus mentis tuae arborem quandam [Draw therefore a similar tree in your mind]," with the detail he himself has just so carefully set out. Modern scholars would expect – and many have – that any paintings or murals made from these words would look more or less the same, and if

[53] Bonaventura, *Lignum vitae*, Prol. 2–3: "in nobis accendatur affectus, formetur cogitatus, imprimatur memoria … et quoniam imaginatio iuvat intelligentiam, ideo quae ex multis pauca collegi in imaginaria quadam arbore sic ordinavi et disposui." I have used the Quaracchi edition, vol. 8, *Opuscula varia ad theologiam mysticam*, ed. A. Lauer, 1898, pp. 68–88. This is digitized (1995) and available online at www.archive.org.

a painter were so unfortunate as to deviate from Bonaventure's instructions, he should be judged as wrong and in need of correction. But Bonaventure emphasizes, rather awkwardly in this sentence, that one is to draw the tree *in spiritus mentis tuis*, not drawing it with or from any external aid or source. He emphasizes "your" – *mentis tuis* – and in that way it will be *your* action, your desire and feelings, and your own spirit's meditation.

The history of these Tree of Life paintings (and many were commissioned, especially in fraternal communities from the late thirteenth century well into the sixteenth) shows great variety among them.[54] The form was soon incorporated into the Franciscan-produced *Speculum theologie*, the popular anthology of invention structures for use by preachers and chaplains alike, compiled in the mid-fourteenth century.[55] Bonaventure's imagined Tree, like his imagined Itinerary, provides a rationally articulated form for use by someone seeking to think about or preach about the Passion for a variety of purposes, private and public, individual and communal, contemplative and pedagogical. Perhaps some of Aristotle's logical distinctions among Causes of things might be helpful in understanding this variety: the painting could be considered as the material cause; the tree or itinerary the formal cause; this form can be used (the instrumental cause) for a variety of persuasive means to a desired goal depending on the occasion (the final cause). I realize that this analysis will seem unduly cumbersome to some readers, but keeping these terms in mind may prevent a student from confusing useful guidance intended for someone's own contemplative thinking with mandatory instructions given by a master to apprentice painters, who thus can get the result truly "wrong" in a necessary sense.

One of the more peculiar iterations of a linear geometric shape for meditative withdrawal and enclosure is the building known as the Rushton Triangular Lodge in Kettering, Northamptonshire (Figure 22). Completed in 1597, it is

54 Sandler 2020 and Esler 2015 have written about Tree of Life paintings that clearly are based on Bonaventure's treatise but render its verbal descriptions loosely and variously. Esler discusses Italian renditions of the Tree from the fourteenth and fifteenth centuries painted on panels, Sandler those in the earliest manuscripts of Bonaventure's text. Perhaps the most famous is the mural by Taddeo Gaddi in the Franciscan church of Sta. Croce, Firenze (1340). Sandler lists the five earliest (thirteenth–early fourteenth century) that are in manuscripts, all conjoined to the verbal text, in p. 445 n9. The manuscript depictions of Hildegard of Bingen's visionary figures are similarly various (and over time have continued to be so, as there is no "correct" rendition). The variance demonstrates clearly that, in monastic contemplation, it is the enlightening *journey* of a particular human mind upwards toward seeing God that matters; for of course, if one has had a true experience of *raptus*, one cannot recall or recount it in human language afterwards anyway.

55 On this popular anthology, see Sandler 1999.

the work of a recusant Catholic landowner, Thomas Tresham, living in Royalist territory during a particularly fierce period of the English pre-Civil War Reformation: James I's Oath of Allegiance was demanded of Catholics only seven years later. Tresham said it served as a lodging for the estate's rabbit keeper (and such estates did breed rabbits for their use at this time).[56] Its three floors each contain a hexagonal room with triangular corners; a winding inner stair up one of these corners gives access. The external walls are solid brick, except for many small triangular openings set in circular or rhomboidal moldings, plus various enigmatic verbal riddles and numbers all invoking in one way or another the number three. This architectural curiosity is usually explained as announcing the name of its owner by punning lavishly on the name Tresham, and Thomas was familiarly called "Tres." But none of these explanations accounts adequately for the building's shape, which is best understood from above, a perspective unavailable to any passerby of the early seventeenth century, though of course visible to God. The Rushton Lodge is the "enigma" of a faith seen "darkly as in a mirror" and materialized in architecture; everything about it shuts it away from the world, while its curious shape and external decoration also loudly call the passing world's attention to itself (though only to a point).

Since it is a solid, the Lodge's triangular shape is that of a prism. Triangularity demonstrates in a way that humans can readily grasp how it is possible to be both three and one at the same time. Moreover, the observation that a prism breaks white light into many colors also made it particularly attractive as a demonstration of how One creates Many, a property analyzed at length during the twelfth and early thirteenth centuries by Robert Grosseteste.[57] In the traditions of Latin hexaemeral analysis, triangles are the linear stamp of the creating Trinity, the Earth's basic measuring form, a position the shape is given (an equilateral triangle formed from two intersecting circles) also in Euclid's First Proposition, the one that logically starts off the rest.

The Triangular Lodge invites and excludes simultaneously. In the best traditions of contemplation, one is secreted from the world, enclosed bodily

[56] The building is now in the keeping of English Heritage, which has an informative history, several photos, including of the Lodge taken from above so that its triangular shape is evident, and also architectural plans of its interior; all these are accessible online at www.english-heritage.org.uk.

[57] Grosseteste's treatise "On Light" (*De luce*) has recently (2013) been re-edited and translated, by C. Panti and with full introductions and notes. His *De iride* ("On the Rainbow") was translated by David C. Lindberg in 1974; a new edition and translation by Sigbjørn Sønnesyn is forthcoming in the series *The Scientific Works of Robert Grosseteste* (Oxford University Press, in progress). See also Smithson et al. 2014.

within a small space and yet enabled in mind to soar as far as one wishes, as Peter of Celle counseled. As a chamber for invention the Lodge is an apt *cella*, since, while its spaces are constrained, their purpose is to enable those within to fly upward in a state of exalted mind in order to see All. It has been suggested that the Lodge was used for religious gatherings of Catholics hiding from the king: if so, that too is consistent with its apparent show and yet complete inner secrecy. As a home for the estate's warrener, its designation (as the "Warreyne") is fully in keeping with some late traditions of Catholic prayer, for the margins of late English prayer-books often depict rabbits in many guises, chasing and being chased around the texts (as in Figure 11), alluding to a common trope of meditative prayer as a hunt following the *vestigia*, the "footprints," of desired hermeneutic meanings in Scripture and liturgy, to uncover the "prey," with a satisfying pun possible in English on "pray."[58]

Some Inventive Shape-Shifting

I would now like to explore some examples of how a particular subject content, shaped in several ways, yields different, though compatible, ways of thinking, a logic of Both/And complexity, rather than the simple Either/Or of the Square of Opposition. The possibilities opened up by plotting arguments using structural re-formations were very well known and exploited both for teaching and in meditation but have not always been remarked as fully and fruitfully as they deserve to be. To return to the arresting analogy that Richard of St. Victor drew between flocking birds and thinking minds, we need to recall that the shapes formed by the birds change frequently. These flocks do not form only one stable shape (like the congruent lines of cranes), but turn and regroup in another shape and yet another as their formations constantly vary. So too the organizing shapes used for pondering the questions of a particular subject matter are varied, not so much in order to find the ultimately best one (for in a complicated many-folded matter, there likely is no single "best"), but to keep a thinking mind from becoming overly repetitive in its habits and thus bored and dull. The great temptation of monastic meditating was to succumb to tedium, that deadening of mind, spirit, and feeling which the daily routine of Offices and hours of Silence was all too apt to induce, and which monastic tradition (alluding to Ps. 90: 6) called "the noon-day devil" (*daemonium meridianum*). The "silence" of contemplative prayer was of necessity kept varied and lively, and its many

[58] I discussed this trope and other related ones in relation to the disciplines of prayerful meditation in Carruthers 1998, ch. 3 esp.

mental (and physical) exercises are intended to address this requirement and help to focus a monk's creating powers.[59]

The Heavenly City itself took on two basic forms, one walled within a quadrangle and the other within a circle. In both shapes twelve towers rise above the walls, just as John described. As principles of meditating, these two basic visualized shapes enable and emphasize different relations of historical time to Eternity. The quadrangle emphasizes the Heavenly City as the New Jerusalem, its form envisioned literally from John's Apocalypse text (as in Figures 19–20). But the walls of Jerusalem are not always shown as square. There are depictions of the idealized City on a Hill as round also, as it is shown in a twelfth-century Danish Cistercian manuscript now in the library of Uppsala University (Figure 23). The circular form emphasizes its perfected "fullness," as the *New* Jerusalem made perfect in Eternity. This is also the shape of German Romanesque wheel chandeliers or coronas, such as those raised in the Palatine Chapel in Aachen and in Hildesheim (especially the "Heziloleuchter," named for the eleventh-century bishop who commissioned it).

The useful rational aptness of this traditional indeterminacy was not confined to medieval artists. It was recognized as late as John Milton. Satan, winging his dark, evil way through Chaos to scout out God's newly formed, as-yet innocent universe, pauses "at leisure to behold/ Far off the empyreal heaven, extended wide/ In circuit, undetermined square or round" (*Paradise Lost*, II.1046–48). Divinity cannot ever be bound in the *formae* of human geometry, the insight that Hildegard's followers also showed when painting her undetermined architectural envisionings, like that of the City of the Church.[60] Acquiring a mental portfolio of applicable, useful shapes offers a most useful toolbox for inventively associating various materials within the library of all one has learned.

[59] The caution applies as well to all composition: as Geoffrey of Vinsauf observed in his *Poetria nova* (Bk. V, 1969–86), by its nature human memory makes for itself "a room of delights," craving what pleases it, not what is boring. It is thus easily distracted unless it is constantly delighted; it also needs to be fed moderately, in digestible chunks. The lively nature of monastic Silence as *otium negotium*, is well characterized by Gehl 1987. On monastic understanding of such periods of "busy leisure," see Leclercq 1963. The vice of tedium (*acedia*) and its many temptations is described, in imaginatively lively fashion, in chs. 12, 27–29 of Evagrius Ponticus's *Praktikos*, a basic text of Desert monastic meditational practice.

[60] These depictions have been well studied and often reproduced, including online and in the translation of *Scivias* (1990). Among the best studies of their geometries are those of Caviness 1988 and Fassler's 2014 and 2022 studies of Hildegard's "allegorical architecture" and her music. Though not specifically about Hildegard, Newman 2013 is an excellent study of such convent-made visionary books.

Diagrams of the Liberal Arts

In the early twelfth century, the scholar Honorius Augustodunensis wrote a treatise *De artibus* in which he imagines the progress of a human soul from its exile in Ignorance to its homeland (*patria*) in Wisdom (*sapientia*) as an itinerary through seven cities (*civitates*) from Babylon to Jerusalem. He describes this also as a gradual journey from full darkness (Babylon) to full light (Jerusalem). On this route the first city is Grammar, where Donatus and Priscian are teaching. The gateway to this city is through three sorts of sounds – vowels, semivowels, and consonants (note that these grammatical sounds are not yet written as *litterae* – in the home of Grammar one *hears* before seeing). These sounds lead in turn to the houses of sentences via the paths of syllables and then whole words, and finally on to the grand villas of the Poetical works (*libri poetarum*), including Lucan, Terence, Persius, and Horace. And in this manner the treatise proceeds. After leaving the city of Grammar one comes to the city of Rhetoric (Cicero teaches there); thence to Dialectic, which is a walled citadel encircled by nine towers (the topical *circumstantiae*, literally "standing around" the citadel of reasoning), in which Aristotle expounds his *Topics*. Thence to Arithmetic, and to Music: both are cities in which Boethius teaches. Next one journeys into Geometry, where, with Aratus's aid, *Geographia* (the world map, *mappa mundi*) is explained, and thence to Astronomy, which has two teachers, "Hyginus," who, with an astrolabe, teaches the solar and planetary motions; and "Julius," who, by means of numerical *computus*, shows how the world's Ages from Creation have unfolded, for Motion is always in space-time, forming the divinely ordained combination of Physics with History.[61]

There are actually ten *artes* in Honorius's route, something not unusual in twelfth-century schemes. The final three are *Physica* (which to Honorius meant Medicine, presided over by Hippocrates); *Mechanica*, all arts made with human hands whether in metal or wood or marble; and *Oeconomica*,

61 "Hyginus" was initially the second-century author of a treatise on the laying out of military camps (*castra*), but whose work in fragmentary form became anthologized as a guide to surveying methods in *Aratus latinus*, the basic text for teaching *Astronomia* in the eleventh- to twelfth-century liberal arts curriculum. The *astrolabium* that Honorius supposes him to have used to survey celestial movement is presumably a pointing tool like those supplied by Gerbert of Aurillac for teaching *astronomia* in Rheims. "Julius" is Julius Africanus, the third-century author of a *Chronographia*, also included in fragments in "the Aratus," which taught not only celestial movement but also how Christian history has unfolded in time from Creation to Resurrection. On the authors found in *Aratus latinus*, see Wallis 2004, esp. pp. lxxxix–xcii, Burnett 1997, and Riché 1989.

what we now would call the Social Sciences, especially Politics and Law, which organize humans communally through the institutional "households" (*economiae*) of Church and State.

I have never seen Honorius's itinerary drawn in a manuscript, but it is nonetheless evidently a diagram-map intended to be imagined as such from his words. The productive dimensions in any itinerary diagram are *temporal* more than spatial, as is true in all medieval maps, for earthly distances were measured in journey times (days and months), not in units of linear space. A medieval Christian *mappa mundi* incorporates two viewpoints simultaneously, one the Majestas Domini, the divine view literally outside of Time but surrounding a view of human salvation History mapped onto a geographical itinerary spreading out from its start in the original expulsion from Eden.[62] Circular diagrams of the cosmos all show the celestial movements in time rather than space, as they measure the daily, seasonal, and annual celestial motions measured of necessity from our human perspective from Earth.

The Arts were also drawn as a Tower arranged in rooms rising in storeys (*gradus*), built up solidly from the foundation of Grammar. Figure 24 reproduces a woodcut from the early sixteenth century, but the curriculum it models is still recognizably that of Honorius, even to some of the great teacher figures who hold the iconic keys to each of its rooms. Each scheme – itinerary and tower – involves movement, and both follow a strict *ductus*: the student cannot skip a grade of the prescribed path. Honorius's itinerary moves horizontally: one enters and passes through each city, although taking time to explore each one as fully as one can, for this is a journey with delays and some stop-overs, in order to learn thoroughly what is taught in each city. But the *Margarita philosophica* tower requires us also to move in a vertical spiral up the stairs and around each floor (*gradus*), the pupil given his basic hornbook of ABCs and prayers,[63] carrying which he can enter the door into Grammar on the ground floor (with two chambers, Donatus in the lower, Priscian in the upper). After learning from them, he mounts the graded stairs to the first floor, circles through Logic, Rhetoric, and Arithmetic: notice that Arithmetic is in the same level as Logic and Rhetoric. Then a student climbs up to Music,

[62] Kupfer 2016 persuasively demonstrates the dual perspective of the wall-sized fourteenth-century Hereford Map, mapping both earth-centered topography and temporal history, but also showing these as reversed from the perspective of the divine Majesty looking down from Eternity. The labels of this *mappa mundi* reverse the temporal-geographical placement of the continents, thereby envisioning both perspectives at once, thereby seeing their relationship as in a mirror, *per speculum*.

[63] Children learned elementary reading at home; see the magisterial account by Clanchy 2018.

Geometry, Astronomy; up another flight to "Physic" (which by the sixteenth century meant *Philosophia naturalis*, the domain of "the Philosopher," that is Aristotle's *Physics* augmented by its many commentaries, Avicenna, Averroes, and some medieval commentators). Moral Philosophy, presided over by Seneca, is the second chamber of this grade. "Theologia seu Metaphysica" is in the very top of the Tower, and its single chamber houses Peter Lombard. But these higher subjects all rest firmly on Grammar – that foundation enables the whole building. To underscore this, the entire education structure is labelled *Typus Grammaticus*, even as it rises to the most speculative of sciences; Grammar is not just abandoned in the chambers at the bottom as the student is promoted (a verb literally meaning "moved up") to the top. And the student's climbing motion is a spiral, an idea of learning as creative, not just a trudge through some old textbooks.

Figure 25 also shows the *divisiones Philosophiae*, but this time as a circle, or rather a circular building with a dome and an arcade, a round-church type of structure, whose prototype is the Holy Sepulcher in Jerusalem. This picture of the Liberal Arts was in a twelfth-century manuscript, *Hortus deliciarum* or "Garden of Pleasures," made by the nuns of Hohenbourg and their abbess, Herrad.[64] We are located just above the top of the dome, looking down with the birds-eye perspective proper to Philosophy. The movement this geometry implies is very different from that of an itinerary or a tower. Instead this diagram explicates what Martianus Capella has Lady Geometry say about all spherical bodies (*sphaerae*): that a sphere, capaciously "containing all other figures within itself, consists of circles into which it is resolved."[65] In this picture Philosophy's sphere resolves into the surrounding arcade that contains each particular Liberal Art within her all-encompassing circle. They are not isolated one from another in rooms but form a linked rotundity. There is also no evident hierarchy implied among them, as inevitably is implied in any linear structure composed of grades and steps, such as a Tower. Instead the Arts are defined as seven "rivers" (*septem fontes*) flowing from Philosophy,

64 *Hortus deliciarum* was an anthology of various materials, put together mostly in the later twelfth century, but with material added. It was in the library of Strasbourg cathedral, and destroyed by fire in 1870, so its images are now reproduced from transcripts and drawings traced from the manuscript in the nineteenth century. See Joyner 2016, *passim*, but esp. pp. 69–76, which give a good account of this particular diagram.

65 Martianus Capella, *De nuptiis*, VI.722: "sphaera sane intrinsecus capax omnium circulis subsistit, in quos resolvitur"; trans. Stahl. Martianus Capella was another author excerpted in the *Aratus latinus*, as was Macrobius. The sphere was considered to be the primary created body in Platonist geometry: see *Timeaus* 55b.

who is their common source and author. This picture also does not have an obvious *principium* among the Arts, only their implied beginning (and ending) in *Philosophia* herself.

A similar circular structure drawn to depict the Liberal Arts is in a manuscript now in the Newberry Library in Chicago (Figure 26). This drawing has no human figures, only summary texts written within its compass-drawn circles, but its layout is strikingly similar to that in the *Hortus*. It was in the library of the Benedictine monastery of Admont until it came to Chicago in 1938; it was most likely also made at Admont around 1100.[66] The texts are summaries of Martianus Capella's descriptions of Philosophy and each of the seven arts. Again, no hierarchy nor beginning is obviously indicated, though in order to read the words in it, the wheel must be imagined as turning clockwise. The Art at the apparent top of the arcade is Astronomy, with Ptolemy as her teacher-figure; then Dialectic (with Aristotle). Next comes Rhetoric (with Tullius) and then Grammar (at the very bottom, with Priscian), just before Music (with Pythagoras, not Boethius). She is followed by Arithmetica, with Nichomachus, teacher of numbers, and then Geometria – the compartment next to her in the arcade although labeled "Euclides" has no text.[67] The logic of the Newberry diagram is evidently different from that of Tower or Journey diagrams showing the relationships of the Arts to Philosophy, but there is no suggestion here either that one type of structure is better or truer than the others. What is clear in each one is that moving through the wheel causes us to view the arts' internal relationships a bit differently and as a result, as we take their measure in themselves tutored by each great teacher, we can also compare each with other, and can understand and value them more comprehensively within what forms a curricular whole. That is, at least, the ideal route of this broadly based and carefully staged education.

[66] The Newberry's catalogue description of this manuscript, now Newberry Library Case MS 9, suggests dates from 1050 to 1150. The chief content of the manuscript is Boethius, *De institutione musica*; there is also a tonary and a brief excerpt from Isidore of Seville's definitions of musical terms (*Etymol.* III.20–1). The drawing is on the verso of the last folio of the book, evidently made somewhat later than the previous folios (closer to 1150). There are no other drawings. Masi 1972 discusses the diagram and transcribes the texts in the inner circle and arcade.

[67] The layout and texts chosen for this diagram on fol. 65v are unusual among liberal arts diagrams. Arithmetic and Music more often are taught from Boethius, whose *De institutione arithmetica* (which included his discussion of Geometry) was regarded as prerequisite to his *De institutione musica*, the main text in this manuscript. Euclid is associated with Geometry in later centuries, as in the "Margarita philosophiae" diagram (Figure 24), but Honorius, more contemporary with the Newberry manuscript, associates her with Aratus and maps. This diagram could use more study by historians of science, education, and art.

Coda

What we call the beginning is often the end
And to make an end is to make a beginning.

(T. S. Eliot, "Little Gidding")

It seems to me that to finish in any definitive manner a book that is about Beginning, with a neat wrap-up in which the great detective Explains All, or that other classic nineteenth-century linear plot in which the characters' later lives are each summed up would be perverse. True lines in the geometry this book describes do not end, but explicate outward infinitely from their creating dimensionless point, often "kissing" others to form the angles that shape all the specific forms in Nature.

So instead of a formal Conclusion, I will end this study by examining a work that in many ways mirrors, though imperfectly, the one I began with, the fifteenth-century Cardinal Nicolas of Cusa's "On a bowling game" (*De ludo globi*). Nicolas used geometry as a dynamic structuring device to teach some important virtues of governance, both of self and state, to young princes of a royal household. The work I want to examine now is also by a cleric teaching novices. But this short work is Victorine, from the later twelfth century, and entirely monastic in its intended audience. It addresses not the ways of secular rulers but the ways of contemplative composing and preaching. It is an *ars predicandi* of a sort but intended for a particular audience, the Augustinian community of St. Victor in Paris, located a short distance away from the precincts of the University that was then just coming into being, although in spirit and institutional mission they seemed far apart.

This Victorine pedagogical *ars* is not well known in scholarship though it has sometimes been published as the conclusion to Richard of St. Victor's "On the Mystical Ark" (*De contemplatione maior*).[1] The work is found in

[1] The text is in *PL* 196.191D–202B. It was translated into English from the PL text by Zinn 1979, pp. 344–70. As Zinn comments, "it offers a great deal of material that is drawn from the longer treatise and compressed into an interesting format"

this appendix-like position in only a few manuscripts of the longer work. It is thus not clearly the work of Richard himself, and no edition, even the seventeenth-century one which Migne published in the *Patrologia Latina*, has claimed it definitely to be so. The latest editor of *De contemplatione*, Jean Grosfillier, does not include it. But whoever composed it was entirely familiar with Richard's longer treatment of his method of contemplative praying, and for the purposes of my examination here the question of who initially wrote it is immaterial. For it is the method employed in this short work that is its most interesting aspect, since it so clearly uses those tools of Topically conceived and Rhetorically intended Invention and Disposition, whose roots are in the late Roman insistence that the Liberal Arts, language and mathematics both, form a curricular whole. More specifically, it makes use of several exemplary meditations expanded from the *geometricalia* of Scripture that I have focused on in the preceding chapters. This particular work demonstrates with unusual flexibility how the inventive potential of logical *Topica* as a powerful *ars inveniendi* as well as *iudicandi* can be directed to much wider and loftier kinds of knowledge, even of salvation, when built like a chariot upon the curricular wheels (*circuli*, *rotae*) of elementary learning.

The treatise's main title is *Some Allegories of the Tabernacle of the Covenant* (*Nonnullae allegoriae tabernaculi foederis*), promising a "very brief recapitulation of the preceding work concerning the grace of contemplation."[2] It becomes clear that it is intended as an *ars inveniendi*, presented in the Cicero-counseled, Boethian-inflected method of medieval *Topica*, intended to show how Richard's method applies not only to individual understanding, but to using the grace of contemplation to teach and persuade others during the many occasions for communal reflection that were a vital part of Victorine life. The final goal of studious meditation is to be able promptly to "prove truth and disprove falsehood; persuade to just action and dissuade from unjust action [In promptu habeant probare vera, improbare falsa, suadere justa, dissuadere injusta]".[3] *Probare* is the combination of Dialectic and *argumenta* with the *suasia* of Rhetoric that produces *confidentia.* Unlike the logically knotted, dense arguments of the academics, the Victorines are always ready and direct ("*in promptu*") with their persuasive proofs.

(Zinn 1979, p. 44). He connects this format to the mnemonic "chunking" method for inventorying and recalling the Psalms that is described by Hugh of St. Victor (*ibid.*, p. 44 n1). I was indebted to Zinn's work on Hugh's art of memory while researching my own *Book of Memory*, in which I discussed and translated Hugh's work; see Carruthers 2008, Appendix A (also Carruthers and Ziolkowski 2002, selection 1).

2 Zinn 1979, p. 344.

3 *Ibid.* p. 367, translating PL 196.200D.

It has been assumed by too many historians that an impassible gulf existed between the intellectual values and motives of contemplative orders like the Carthusians, Augustinians, and Cistercians and the rationalist (and increasingly influential) Aristotelians who taught inside the Parisian city walls. Hugh and other Victorines complain of the noisy debates of preening academics (those ever-useful targets) whom Peter of Celle called the "superficial and jabbering disputants" who publically debated their "unnecessary" questions.[4] Peter contrasts these in motive, method, and outcome to the peace and fruitful quiet of his own *cella* where, in his meditations, he can explore the one great question, seeking oneself in God and God in oneself, "se in Deum et Deum in se quaerere."[5] In the first part of *De contemplatione maior*, Richard recognizes an apparent logical binary of reason and faith in terms of what he calls the *officina Aristotelis*, "the workshop of Aristotle" (the human knowing shaped within the *Organon*, culminating with the *Topica*, which incorporates Logic's necessary mathematical companions), and the *officina Salvatoris*, "the workshop of the Saviour," that persuades and raises the reasoning mind (*elevatio mentis*) to true *fides*, the virtue of *sapientia*, wisdom supported in confident, loving belief.

Some Allegories makes clear through its method of interpreting Richard's great work that *officina Aristotelis* and *officina Salvatoris* should not be understood only on a simple Square of Opposition, though the Square is a secure place to begin. The Saviour's workshop supersedes ("rises over") Aristotle's but the methods of measuring learned in the study of Logica are what enable (strengthened by grace) the reasoning mind to rise towards the Saviour. *Some Allegories* calls this *gratia discretionis* ("the gift of discerning," analyzing and judging truly) and *studiosa meditatio* ("studious meditating"). This is a typically Victorine idea, explored fully in Hugh's greatly influential *Didascalicon*, in the Carthusian abbot Guigo II's *Scala paradisi*, and in John of Salisbury's *Metalogicon*. Anselm of Bec was a fountain for them all.[6]

Some Allegories links the gift of discernment to the divinely ordained Candelabrum placed in the outer chamber of the Tabernacle structure, which sheds its true light to elevate human understanding: it is also an expression of the Augustinian concept of the divinely granted "inner light" of the mind,

4 Peter of Celle, *De afflictione et lectione*, ed. Leclercq 1946, p. 238.22–23 (trans. Feiss 1987, p. 140).

5 *Ibid.* p. 238.15.

6 Anselm's role in the twelfth-century Augustinian revival, marrying logic and prayer, has been magisterially studied by Southern 1990 and Ward 2009.

lucerna interioris hominis.[7] This humanly available *lucerna* is not always used well, for it can be perverted or ignored, but it is a salvific light, making the way possible for people to rationally lay out their path to truthful thought and confident belief. In Richard's terms, all contemplation must start in the manner of Bezelel, the architect who materially envisioned and constructed the Tabernacle and Ark according to God's initial instructions to Moses. Only a very rare few are then granted the "supervening" grace given to Aaron and admission to the Holy of Holies itself, and only then after having mastered the human way of Bezelel.

Since God's direct speaking is veiled in a cloud and far beyond any human language, Moses first had to translate the divine plans for Bezelel (and for Aaron) into language he could understand, and so be able faithfully to ordain, build, and instruct human society. And the language that Moses primarily used for this mediating task was Mathematics (Arithmetic, Geometry, and Music), the measures of the saving Covenant between Divine and human, expressed architecturally through the numbered dimensions of the Tabernacle and persuasively communicated through Grammar, Dialectic, and Rhetoric.[8]

Geometrical reasoning dominates the method propounded in *Some Allegories* for envisioning the Ark. As the author (Richard?) focuses on its significations (the Ark itself is placed in the inner chamber of the Tabernacle structure), his contemplative method becomes more overtly geometrical and dialectical as he attempts to elevate and expand his thinking into the "free gaze of the mind suspended with wonder in the Showings of wisdom [Contemplatio autem est libera mentis perspicacia in sapientiae spectacula cum admiratione suspensa]," his definition of contemplation.[9]

My several examinations of medieval compositions employing such wondrous varieties of geometric shapes are not intended to be in any way definitive, but to open up some possibilities and invite readers to make their own adaptations

7 *Nonnullae allegoriae*, PL 196.193. Much has been written about this Augustinian concept; see Nash 1969, and in the specifically Cistercian context, McGinn 1977, Mews 2018. In Carruthers 2025, I discussed this three-fold Ricardian path (Moses, Aaron, Bezalel) in greater detail.

8 The relevant text is Exodus 25–27, where the Tabernacle's furnishings and dimensions are set out in great detail. The sounding of the High Priest's horn (Music), though not commented on in *Some Allegories*, is essential to expressing the divine Covenant as well. It was not difficult for Judeo-Christian exegetes to map the Exodus text onto the traditional Liberal Arts, as Cassiodorus demonstrated.

9 PL 196.193; I am using Julian of Norwich's English equivalent of *spectacula*. Hugh of St. Victor makes a similar distinction between the disciplines of *lectio* and of *meditatio* in *Didascalicon* III, while also arguing that they build upon one another.

and translations of these fruitful structures for many different meditations and inventions. I said at first that this book is about beginning. It seems appropriate to end it with an invitation for its readers to start upon enjoyable, beneficial, and persuasive inventions of their own. This has certainly been the traditional reason for those studying the crafts of language to make also careful study of geometry and of numbers. That *mens a mensura dicitur* is a fundamental belief of medieval rhetoric and all rational creative expression.

In 1570 the first translation of Euclid into English was published in London, together with an extensive *Mathematicall Preface* by the Elizabethan polymath, Dr. John Dee. Geometry, he says, is not just a matter of "creping on ground … with pole-perche, rod or lyne: but liftyng the hart aboue the heauens, by inuisible lines, and immortall beames: [it] meteth with the reflexions, of the light incomprehensible, and so procureth Ioye and perfection vnspeakable.… Geometrie is the knowledge of that which is euerlastyng. It will lift vp therfore (O gentle Sir) our mynde to the Veritie: and by that meanes, it will prepare the Thought to the Philosophicall loue of wisdome: that we may turne or convert toward heauenly thinges, both mynde and thought."[10]

10 Dee, *Mathematicall Praeface*, a.iii–iv.

Bibliography

Primary Sources

Adam Scot. "De libro atque pictura tripartiti tabernaculi." PL 198. 609–794.

Adam Scot [Guigo II]. "Liber de quadripartito exercitio de cellae." PL 153. 800–84.

Alan of Lille. *Literary Works*, edited with translations by W. P. Wetherbee. DOML 22. Cambridge, MA: Harvard University Press, 2013.

Aristotle. *On the Soul (De Anima), Parva Naturalia, On Breath*, translated by W. S. Hett. LCL 288. Cambridge, MA: Harvard University Press, 1936.

——. *Metaphysics, Books 1–9*, translated by H. Tredennick. LCL 271. Cambridge, MA: Harvard University Press, 1933.

Augustine of Hippo (Augustinus Hipponensis). *De Genesi ad litteram libri duodecem*, edited by J. Zycha. CSEL 28. Vienna: Tempsky, 1894. Repub. in CCSL, online in LLT.

——. *On Genesis*, translated by E. Hill. Hyde Park: New City Press, 2002.

——. *The Literal Meaning of Genesis*, translated by J. H. Taylor. Ancient Christian Writers, 41–2. New York: Newman Press, 1982.

——. *De Trinitate*, edited by W. J. Mountain. CCSL 50. Turnhout: Brepols, 1968. Translated as *The Trinity* by E. Hill. Hyde Park: New City Press, 1991.

——. *De Ordine*, edited by W. M. Green. CCSL 29. Turnhout: Brepols, 1970. Translated as "On Order," in M. P. Foley, *St. Augustine's Cassiciacum Dialogues*, vol. 3. New Haven: Yale University Press, 2020.

——. *De Quantitate Animae*. PL 32.1035–80; CSEL 89 (ed. Hormann, 1986). Translated as *The Magnitude of the Soul* by J. McMahon. Fathers of the Church 4. Washington: Catholic University Press, 1947.

——. *Confessionum Libri XIII*, edited by L. Verheijen. CCSL 37. Turnhout: Brepols, 1990.

——. *Confessions*, edited and translated by J. J. O'Donnell. 3 vols. Oxford: Clarendon Press, 1992.

——. *On Christian Teaching*, translated by R. P. H. Green. Oxford: Oxford University Press, 1997.

——. *Opera Omnia. CAG Online* (Corpus Augustinianum Gissense), edited by C. Mayer. In "Past Masters" Database. Charlottesville: InteLex, 2000, by library subscription.

——. *De Doctrina Christiana*, edited by J. Martin. CCSL 32. Turnhout: Brepols, 1962.

——. *Soliloquiorum libri duo.* PL 32, cols. 869–904.

Bacon, Roger. "Mathematicae in divinis utilitas." In J. H. Bridges, ed. *Opus maius*, 3 vols. Vol. 1, 175–402. Oxford: Clarendon Press, 1897. (Online corrected edition, LLT, accessed February 23, 2023).

——. *Opus Tertium* Teil I, ch 58; In Nikolaus Egel, ed. and transl. (Latin and German): 472–7. Philosophische Bibliothek 718. Hamburg: Felix Meiner Verlag, 2020. (Also ed. J. S. Brewster (1859), available online in LLT, accessed March 16, 2023.)

Bede. *De temporum ratione*, edited and translated by Faith Wallis in *Bede: The Reckoning of Time*. Liverpool: Liverpool University Press, 2004.

Benedict of Nursia. *The Rule of St. Benedict in Latin and English with Notes.* Collegeville, MN: The Liturgical Press, 1980.

Bernard of Clairvaux. *Sancti Bernardi Opera*, edited by C. H. Talbot, J. Leclercq, and H. M. Rochais. 6 vols. Rome: Editiones Cistercienses, 1957–74.

Bernardus Silvestris. *Collected Works*, edited by W. P. Weatherbee. DOML 38. Cambridge, MA: Harvard University Press, 2015.

——. *The Cosmographia of Bernardus Sylvestris*, edited and translated by, W. P. Weatherbee. New York: Columbia University Press, 1973.

Boethius, A. M. S. *De topicis differentiis*, edited by E. Stump. Ithaca: Cornell University Press, 1978.

——. *Consolatio Philosophiae*, edited by L. Bieler.CCSL 94. Turnhout: Brepols, 1984.

——. *In Ciceronis Topica*, edited by E. Stump. Ithaca: Cornell University Press, 1988.

——. "De trinitate." In *Theological tractates*, edited and translated by H. F Stewart, E. K. Rand, and S. J. Tester, 2–31. LCL 74. Cambridge, MA: Harvard University Press, 1973.

——. *De Institutione Arithmetica; De Institutione Musica, accedit Geometria quae fertur Boetii*, edited by G. Friedlein. Leipzig: Teubner, 1867.

——. *Fundamentals of Music (De Musica)*, translated by C. M. Bower. New Haven: Yale University Press, 1989.

——. *Boethian Number Theory: A Translation of the De Institutione Arithmetica*, translated by M. Masi. Amsterdam: Rodopi, 1983.

Bonaventura, Saint. *The Soul's Journey into God, the Tree of Life, the Life of St. Francis*, translated by E. Cousins. New York: Paulist Press, 1978.

——. *Doctoris Seraphici S. Bonaventurae … Opera Omnia*. Quaracchi. 10 vols. Rome: Collegium S. Bonaventurae, 1882–1902.

Boncompagno da Signa. *Rhetorica novissima*, edited by A. Gaudenzi. 3 vols. Vol. 2: 249–97. Bologna: Societas Azzoguidiana 1892.

Calcidius. *On Plato's Timaeus*, edited by. John Magee. DOML 41. Cambridge, MA: Harvard University Press, 2016.

Cassiodorus Senator. *Institutiones divinarum et saecularium litterarum*, edited by R. A. B. Mynors. Oxford: Clarendon Press, 1963.

——. *Variae*, edited by T. Mommsen and L. Traube. MGH *auctores antiquissimi*, 1894. Translated by M. S. Bjornlie, *The variae: the complete translation*. Oakland: University of California Press, 2020.

——. *Expositio Psalmorum*, edited by M. Adriaen. 2 vols. CCSL 97–8. Turnhout: Brepols, 1958.

Chastellain, Georges, Jean Robertet and Jean de Montferrant. *Les Douze Dames de Rhetorique*, edited by David Cowling. Geneva: Droz, 2002.

Cicero, M. T. *De inventione. De optimo genere oratorum. Topica*, translated by. H. M. Hubbell. LCL 386. Cambridge, MA: Harvard University Press, 1949.

——. *Pro Rege Deiotaro*. In *Orationes: Pro Milone*, etc., translated by N. H. Watts. LCL 252, pp. 500–41. Cambridge, MA: Harvard University Press, 1931.

Clarembald of Arras. *Life and Works*, edited by N. M. Häring. Toronto: PIMS, 1965. D. B. George and J. R. Fortin, trans., *The Boethian Commentaries of Clarembald of Arras*. Notre Dame, 2002.

Cosmas Indicopleustes. *The Christian Topography of Cosmas, an Egyptian Monk*, translated by J. W. McKrindle. Cambridge Library Collection. Haklyut Society First Series. Orig. publ. London: 1897. Cambridge University Press, 2010. Online at Cambridge Core.

——. *Topographie Chrétienne*, edited by W. Wolska-Conus, with French trans. 3 vols. (SC 141, 159, 197). Paris: Cerf, 1968–73.

D'Alverny, Marie-Thérèse. *Alain De Lisle: Textes Inédits*. Paris: Vrin, 1965.

Dante Alighieri. *La Commedia secondo l'antica vulgata. Le Opere di Dante Alighieri*, edited by G. Petrocchi. Edizione nazionale (4 vols.). Milano: A. Mondadori, 1966–67.

Dee, John. *The Mathematicall Praeface to the Elements of Geometrie of Euclid of Megara*. London: John Day, 1570.

Echardus Magister (Meister Eckhart). *Opus tripartitum: Expositio libri Exodi*, edited by H. Fischer, J. Koch, K. Weiss. In *Meister Eckhart – Die deutschen und lateinischen Werke, Die lateinischen Werke*, Zweiter Band. Stuttgart: Kohlhammer, 1992.

Evagrius Ponticus. *The Praktikos & Chapters on Prayer*, translated by J. E. Bamberger OCSO. Piscataway, NJ: Gorgias Press, 2009.

Geoffrey of Vinsauf. *Poetria nova*. In Edmond Faral, ed. *Les arts poétiques du xiie et du xiiie siècle*. Paris: Champion, 1924. Translated by M. F. Nims (revised M. Camargo). Toronto: PIMS Press, 2010.

Gerald of Wales, *Topographia Hibernica*, vol. 5, edited by J. S. Brewer and J. F. Dimock. Cambridge Library Collection Rolls. Cambridge University Press, 2012. J. J. O'Meara, trans., *The History and Topography of Ireland*, London: Penguin, 1982

Gerbert of Aurillac. "Ysagoge Geometriae." In *Gerberti Opera Mathematica (972–1003)*, edited by N. Bubnov. Hildesheim: Olms, 1963. Also "Geometria Gerberti," edited by. B. Pezius (1853) in PL 139, cols. 91B–152C.

Gerson, Jean. *Oeuvres complètes*, edited by P. Glorieux. 10 vols. Paris, Tournai: Desclée, 1960–73.

Gregory the Great. *Dialogorum libri iv*, edited by A. de Vogüé. SC 260, 265. Paris: Cerf, 1979.

Gregory of Nyssa, *De anima et resurrectione* (1642); rpt PG 46. W. Moore and H. A.Wilson, trans., *Nicene and Post-Nicene Fathers*, Second Series vol. 5, Edinburgh: Clark, 1893.

Grosseteste, Robert. *Hexaemeron*. Oxford: Oxford University Press for the British Academy, 1982. C. F. J. Martin, trans., *On the Six Days of Creation*. Oxford: Oxford University Press for The British Academy, 1996.

——. "Robert Grosseteste's *De Luce*," edited by C. Panti; with N. Lewis, "Robert Grosseteste *De luce* (An English translation)." In J. Flood et al., eds. *Robert Grosseteste and His Intellectual Milieu*, 193–247. Toronto: PIMS, 2013.

——. "Tractatus de fractionibus radiorum," edited by L. Bauer, *Die Philosophischen Werke des Robert Grosseteste* (Muenster: Aschendoff, 1912). Translated by C. Panti, *On Lines, Angles, and Shapes, or On the Refractions and Reflections of Rays* (private communication, quoted with permission).

——. *The Dimensions of Colour: Robert Grosseteste's De Colore*, edited by G. Dinkova-Bruun, G. Gasper, M. Huxtable, T. McLeish, C. Panti, and H. Smithson. Durham Medieval and Renaissance Texts. Toronto: PIMS, 2014.

——. "*De generatione sonorum* (1211): On the Generation of Sounds." In *Knowing and Speaking*, S. O. Sønnesyn, ed. *The Scientific Works of Robert Grosseteste* vol. 1, 243–55. Oxford: Oxford University Press, 2019.

Guigo II. *The Ladder of Monks ("scala Claustralium") and Twelve Meditations ("Meditationes")*, translated by E. Colledge and J. Walsh. Kalamazoo, MI: Cistercian Publications, 1979.

Hildegard of Bingen. *Scivias*, edited by A. Führkötter and A. Carlevaris. CCCM 43. Turnhout: Brepols, 1978. Translated by C. Hart and J. Bishop, New York: Paulist Press, 1990.

——. *Liber Divinorum Operum*, edited by A. Derolez and P. Dronke. CCCM 92. Turnhout: Brepols, 1996. N. M. Campbell, trans., *The Book of Divine Works*. Washington, DC: Catholic University of America, 2018.

Horace. *Satires. Epistles. The Art of Poetry*, edited and translated by H. R. Fairclough. LCL 194. London: Heinemann, 1926.

Hugh of St. Victor. *Didascalicon*, edited by. C. H. Buttimer. Washington, DC: Catholic University Press, 1939. J. Taylor, trans., *Didascalicon*. New York: Columbia University Press, 1961.

Isidore of Seville. *Etymologiarum Sive Originum Libri XX*, edited by R. M. Lindsay. Oxford Classical Texts. 2 vols. Oxford: Oxford University Press,

1911. W. J. Lewis, S. A. Barney et al., trans., *The Etymologies of Isidore of Seville*. Cambridge: Cambridge University Press, 2006.

——. *Traité de la Nature*, edited by Jacques Fontaine. Bordeaux: Féret et Fils, 1960. C. B. Kendall and F. Wallis, trans., *On the Nature of Things (De Natura Rerum)*. Liverpool: Liverpool University Press, 2016.

Jerome. *Sancti Eusebii Hieronymi Epistulae*, edited by I. Hilberg. CSEL 54–56, 4 vols. in 3 (1910–18); rev. M. Kamptner 1996. Vienna: Österreichischen Akademie der Wissenschaften.

John of Salisbury. *Metalogicon*, edited by J. B. Hall and K. S. B. Keats-Rohan. CCCM 98, Turnhout: Brepols, 1991. D. D. McGarry, trans., *The Metalogicon of John of Salisbury: A Twelfth-Century Defense of the Verbal and Logical Arts of the Trivium*. Berkeley: University of California Press, 1955.

John Scottus Eriugena. *Annotationes in Marcianum (Martianum Capellam)*, edited by Cora E. Lutz. Cambridge, MA: Medieval Academy of America, 1939.

Julian of Norwich. *A Book of Showings to the Anchoress Julian of Norwich*, edited by E. College and J. Walsh. Toronto: PIMS, 1978.

Langland, William. *The Vision of Piers Plowman: A Critical Edition of the B-Text*, edited by A. V. C. Schmidt. London: Dent, 2011.

Macrobius. *Commentarii in Somnium Scipionis*, vol. 2, edited by J. Willis. Stuttgart: Teubner, 1979. W. H. Stahl, trans., *Commentary on the Dream of Scipio*. New York: Columbia University Press, 1952.

Martianus Capella, *De nuptiis Philologiae et Mercurii*, edited by J. Willis. Leipzig: Teubner 1983. W. H. Stahl et al., trans., *Martianus Capella and the Seven Liberal Arts*, vol. 2. New York: Columbia University Press, 1971.

Nicolas of Cusa [Cusanus]. *Dialogus de ludo globi.* Paris 1514, re-edited by H. G. Senger. In *Nicolai de Cusa opera omnia*, vol. 9. Hamburg: Heidelberger Akademie der Wissenschaften, 1998. J. Hopkins, trans., *Complete Philosophical and Theological Treatises of Nicholas of Cusa*, vol. 2: 1179–1256. Minneapolis: Banning, 2001. Both are available online at http://www.cusanus-portal.de. P. M. Watts, trans., *De Ludo Globi – the Game of Spheres*. New York: Abaris Books, 1986.

——. *Idiota de mente*. Paris 1514 edition (as above), vol. 1, pp. 48–160. Available online at http://www.cusanus-portal.de.

Peter of Celle. *De afflictione et lectione*, in J. Leclercq, ed. (q.v.), *La spiritualité de Pierre de Celle*, 231–9. H. Feiss, trans., "On Affliction and Reading," *Selected Works of Peter of Celle*: 131–41. Kalamazoo: Cistercian Studies, 1987.

Peter the Chanter. *The Christian at Prayer*, edited by R. C. Trexler. Binghamton, NY: Center for Medieval and Early Renaissance Studies, 1987.

Petrarca, Francesco. *Le familiari*, edited by U. Dotti. 2 vols. Urbino: Argalìa Editore, 1974.

Priscian, *Institutiones grammaticae*, in H. Keil, ed. *Grammatici latini*, vol. 2. Leipzig: Teubner, 1865–80.

Quintilian. *Institutionis Oratoriae Libri Duodecim*, edited by M. Winterbottom. 2 vols. Oxford: Clarendon Press, 1970. D. A. Russell, trans., *The Orator's Education*. 5 vols. LCL 124–7. Cambridge, MA: Harvard University Press, 2001.

Richard de Saint-Victor. *Les Douze Patriarches (Benjamin Minor)*, edited by J. Chatillon and M. Duchet-Suchaux. SC 419. Paris: Editions du Cerf, 1997.

——. *De Contemplatione*, edited by Jean Grosfillier. Turnhout: Brepols, 2013.

——. *The Twelve Patriarchs, The Mystical Ark, Book Three of the Trinity*, translated by Grover A. Zinn. New York: Paulist Press, 1979.

Richer of Saint-Remy. *Richerus Sancti Remigii Remensis monachus, Historiae*, edited by H. Hoffmann, MGH Scriptores, vol. 38. Hanover: Hahnsche, 2000; available online. Translated by Justin Lake, 2 vols. DOML vols. 10–11. Cambridge, MA: Harvard University Press, 2011.

Sedulius Scottus. *Commentum in Majorem Donatum Grammaticum*, edited by D. Brearley. Toronto: PIMS, 1975.

Tartaret, Pierre [=Petrus Tartaretus]. *Expositio super textu logices Aristotelis* (orig. publ. Lyon: Nikolaus Wolf, 1500); available in Early European Books Online.

Thierry of Chartres. *Commentaries on Boethius*, edited by N. M. Häring. Toronto: PIMS, 1971.

Thomas Aquinas. *Summa Theologiae*, online at www.corpusthomisticum.org.

——. *Lectura Romana in Primum Sententiarum Petri Lombardi*. Toronto: PIMS, 2006.

——. *In Psalmos reportatio*, edited by Parmensis, 1863, vol. 14, p. 160. In *Opera Omnia*, online at www.corpusthomisticum.org.

——. *Works in English*. Online at https://isidore.co/aquinas/

Uguccione da Pisa. *Derivationes*, edited by E. Cecchini et al. 2 vols. Florence: SISMEL, Edizioni del Galluzzo, 2004.

Vincent of Beauvais (=Uincentius Beluacensis). *Speculum maius: Speculum doctrinale* (edition of Balthazar Bellère, Douai 1624). Turnhout: Brepols 2020, and online in LLT.

William of Conches. *Dragmaticon Philosophiae*, edited by I. Ronca and A. Badia. CCCM 152. Turnhout: Brepols, 1997.

Secondary Sources

Akbari, Suzanne C. *Seeing through the Veil: Optical Theory and Medieval Allegory*. Toronto: University of Toronto Press, 2004.

Albertson, David. *Mathematical Theologies: Nicholas of Cusa and the Legacy of Thierry of Chartres*. Oxford: Oxford University Press, 2014.

Allen, H. E. "The Fifteenth Century 'Associations of Beasts, of Birds, and of Men'," *PMLA* 51 (1936): 602–6.

Auerbach, Erich. "Figura." R. Mannheim, trans. In P. Valesio, ed. *Scenes from the Drama of European Literature*, 11–71. Minneapolis: University of Minnesota Press, 1984.

Bartlett, Robert. *The Hanged Man: A Story of Miracle, Memory, and Colonialism in the Middle Ages*. Princeton: Princeton University Press, 2006.

Bent, Margaret. "Songs without Music in Dante's *De Vulgari Eloquentia*: *Cantio* and Related Terms." In Teresa M. Gialdroni, et al., eds. *"Et Facciam Dolçi Canti": Studi in Onore di Agostino Ziino*, vol. I, 161–81. Lucca: LIM-Libreria Italiana Musicale, 2004.

Berger, Karol. "The Guidonian Hand." In M. Carruthers and J. M. Ziolkowski, eds. *The Medieval Craft of Memory*, 71–82. Philadelphia: University of Pennsylvania Press, 2002.

Berthoz, Alain. *Le sens du movement*. Paris: Jacob, 1997.

Binski, Paul. "Villard de Honnecourt and Invention." In L. Iseppi De Filippis, ed. *Inventing a Path: Studies in Medieval Rhetoric in Honour of Mary Carruthers*. Nottingham Medieval Studies, 63–79. Turnhout: Brepols, 2012.

——. *Gothic Wonder: Art, Artifice and the Decorated Style 1290–1350*. New Haven: Yale University Press, 2014.

——. *Gothic Sculpture*. New Haven: Yale University Press, 2019.

Binski, Paul, P. N. R. Zutshi, and Stella Panayotova, eds. *Western Illuminated Manuscripts: A Catalogue of the Collection in Cambridge University Library*. Cambridge: Cambridge University Press, 2011.

Black, Deborah L. *Logic and Aristotle's Rhetoric and Poetics in Medieval Arabic Philosophy*. Leiden: Brill, 1990.

——. "Imagination and Estimation: Arabic Paradigms and Western Transformations." *Topoi* 19 (2000): 59–75.

Boitani, Piero. "The Poetry and Poetics of the Creation." In R. Jacoff, ed. *The Cambridge Dante Companion*, rev. ed., 218–35. Cambridge: Cambridge University Press, 2007.

Bolzoni, Lina. *The Web of Images: Vernacular Preaching from Its Origins to St. Bernardino da Siena*. C. Preston and L. Chien, trans. Aldershot: Ashgate, 2004.

Bork, Robert O. *The Geometry of Creation*. Farnham: Ashgate, 2011.

Børtnes, Jostein and Tomas Hägg, eds. *Gregory of Nazianzus Images and Reflections*. Copenhagen: Museum Tusculanum Press, 2006.

Bower, Calvin M. "*Sonus* Vox, Chorda, Nota: Thing, Name, and Sign in Early Medieval Theory." In M. Bernhard, ed. *Quellen Und Studien Zur Musiktheorie Des Mittelalters*, Vol. 3, 47–61. Munich: Bayerische Akademie der Wissenschaften, 2001.

Brown, Peter R. L. *Augustine of Hippo: A Biography*, new ed. Berkeley: University of California, 2000.

——. *Power and Persuasion in Late Antiquity: Towards a Christian Empire*. Madison, WI: University of Wisconsin Press, 1992.

Burnett, Charles. "Sound and Perception in the Middle Ages." In M. Fend, C. Burnett, and P. Gouk, eds. *The Second Sense*, 43–69. London: Warburg Institute, 1991.

——. "The Chapter on the Spirits in the *Pantegni* of Constantine the African." In C. Burnett and D. Jacquart eds. *Constantine the African and 'Ali Ibn Al-'Abbas Al Maǧūsī: The Pantegni and Related Texts*, 99–120. Leiden: Brill, 1994.

——. *The Introduction of Arabic Learning into England*. London: The British Academy, 1997.

——. "*Algorismi vel helcep decentior est diligentia*: the arithmetic of Adelard of Bath and his school." In *Numerals and Arithmetic in the Middle Ages*, 244–52. Variorum Collected Studies. Farnham, Surrey: Ashgate, 2010.

——. "Manuscripts of Latin Translations of Scientific Texts from Arabic." In E. Steiner and L. Ransom, eds. *Taxonomies of Knowledge*, 80–9. Philadelphia: University of Pennsylvania Press, 2015.

Burnyeat, Myles F. "Wittgenstein and Augustine's *De Magistro*." *Proceedings of the Aristotelian Society, Supplement* 61 (1987): 1–24.

——. "Enthymeme: Aristotle on the Rationality of Rhetoric." In A. O. Rorty, ed. *Essays on Aristotle's Rhetoric*, 88–115. Berkeley: University of California Press, 1996.

——. "Postscript on Silent Reading." *Classical Quarterly* 47.1 (1997): 74–76.

Burrus, Virginia. "Is Macrina a Woman? Gregory of Nyssa's *Dialogue on the Soul and Resurrection*." In G. Ward, ed. *The Blackwell Companion to Postmodern Theology*, 249–64. Oxford: Blackwell, 2005.

Bursill-Hall, G. L. *Speculative Grammars of the Middle Ages*. The Hague: Mouton, 1971.

Bynum, Caroline W. "Avoiding the Tyranny of Morphology; or, Why Compare?" *History of Religions* 53.4 (2014): 341–68.

——. *Dissimilar Similitudes: Devotional Objects in Late Medieval Europe*. Princeton: Princeton University Press, 2020.

Cain, Andrew. *The Letters of Jerome: Asceticism, Biblical Exegesis, and the Construction of Christian Authority in Late Antiquity*. Oxford: Oxford Scholarship Online, 2009.

Camille, Michael. "Illuminating Thought: The Trivial Arts in British Library Burney ms. 275." In P. Binski and W. Noel, eds. *New Offerings, Ancient Treasures*, 343–66. Stroud: Sutton, 2001.

Carey, Benedict. "Learning to See the Data." *The New York Times Sunday Review*, March 27, 2015.

Carruthers, Mary. *The Book of Memory. Second edition* (re-issued and revised). Cambridge: Cambridge University Press, 2008.

——. *The Craft of Thought: Rhetoric, Meditation, and the Making of Images, 400–1200*. Cambridge: Cambridge University Press, 1998.

——. "Boncompagno at the Cutting-Edge of Rhetoric: Rhetorical *Memoria* and the Craft of Memory." *The Journal of Medieval Latin* 6 (1996): 44–64.

——. "The Mystery of the Bedchamber: Mnemotechnic and Vision in Chaucer's Book of the Duchess." In J. M. Hill and D. Sinnreich-Levi, eds. *The Rhetorical Poetics of the Middle Ages: Essays in Honor of Robert O. Payne*, 67–87. New Jersey: Fairleigh-Dickinson University Press, 2000.

——. "*Ars oblivionalis, ars inveniendi*: The Cherub Figure and the Art of Memory". *Gesta* 48 (2009): 1–19.

——. "The Concept of *Ductus*, or, Journeying through a Work of Art." In M. Carruthers, ed. *Rhetoric Beyond Words* (q.v.), 190–213.

——, ed. *Rhetoric Beyond Words: Delight and Persuasion in the Arts of the Middle Ages*. Cambridge: Cambridge University Press, 2010.

——. "The 'Pictures' of Jerusalem in Oxford, Bodleian Library MS. Laud Misc. 156." In L. Donkin and H. Vorholt, eds. *Imagining Jerusalem in the Medieval West*, 97–121. Proceedings of the British Academy 175 (2012).

——. *The Experience of Beauty in the Middle Ages.* Oxford-Warburg Studies. Oxford: Oxford University Press, 2013.

——. "Virtue, Intention and the Mind's Eye in Chaucer's *Troilus and Criseyde*." In C. Brewer and B. Windeatt, eds. *Traditions and Innovations in the Study of Middle English Literature (Essays for Derek Brewer)*, 73–87. Woodbridge: Boydell and Brewer, 2013.

——. "*Terribilis, horribilis*, and the fear of God." In N. Jacobs and G. Morgan eds. *"Truthe is the beste": A Festschrift in Honour of A. V. C. Schmidt*, 17–36. Oxford: Peter Lang, 2014.

——. "'The Desert,' Rhetorical Invention, and Prayer in the Augustinian Renewal of the Twelfth Century." In G. de Nie, H. van Loon et al., eds. *Prayer and the Transformation of the Self in Early Christian Mystagogy*, 393–408. Leuven: Peeters, 2018.

——. "Two Unusual Mind Diagrams in a Late Fifteenth-Century Manuscript (UPenn Schoenberg Collection, Ms LJS 429)." *Manuscript Studies: a Journal of the Schoenberg Institute for Manuscript Studies* 4.2 (2019): 389–400.

——. "Geometries for Thinking Creatively." In M. Kupfer et al., eds. *The Visualization of Knowledge in Medieval and Early Modern Europe*, 33–44. Turnhout: Brepols, 2020.

——. "Reading." In M. Gragnolati et al., eds. *The Oxford Handbook of Dante*, Oxford Handbooks Online, 34–48. Oxford: Oxford University Press, 2021.

——. "Of Notes and Unfolding Sounds: Perspectives from Grammar and Geometry." *European Drama and Performance Studies* 19.2 (2022): 59–63.

——. "Encountering Vision: Dislocation, Disquiet, Perplexity in Bonaventure, *The Squire's Tale* and *Pearl*." In A. Butterfield, I. Johnson, and A. Kraebel, eds. *Literary Theory and Criticism in the Later Middle Ages:Interpretation, invention, imagination* (*Essays in Honour of Alastair Minnis*), 263–81. Cambridge: Cambridge University Press, 2023.

——. "Articulating the *geometricalia* of Scripture: Bezelel, Aaron, and Moses". In Maria Streicher and Andrea Worm, eds. *The Menorah and the Seven-branched Candelabrum*. Berlin: De Gruyter, 2025, pp. 79–96.

—— and J. M. Ziolkowski, eds. *The Medieval Craft of Memory: An Anthology of Texts and Pictures*. Philadelphia: University of Pennsylvania Press, 2002.

Caviness, Madeleine H. "Images of Divine Order and the Third Mode of Seeing." *Gesta* 22 (1983): 99–120.

——. "Hildegard as Designer of the Illustrations to Her Works." In C. Burnett and P. Dronke, eds. *Hildegard of Bingen: The Context of Her Thought and Art*, 29–62. Warburg Institute Colloquia. London: Warburg Institute, 1998.

Cerquiglini, Bernard. *Éloge de la variante: histoire critique de la philologie*. Paris: Éditions du Seuil, 1989.

Chaganti, Seeta. *Strange Footing: Poetic Form and Dance in the Late Middle Ages*. Chicago: University of Chicago Press, 2018.

Chajes, J. H. "Spheres, Sefirot, and the Imaginal Astronomical Discourse of Classical Kabbalah." *Harvard Theological Review* 113.2 (2020): 230–62.

Chalmers, David J. *The Conscious Mind: In Search of a Fundamental Theory*. Oxford: Oxford University Press, 1996.

Chen-Morris, Raz. "Geometry and the Making of Utopian Knowledge in Early Modern Europe." *Nuncius* 35 (2020): 387–412.

Chenu, M. D. "Auctor, Actor, Autor." *Bulletin Du Cange* 3 (1927): 81–6.

——. "*Spiritus*: Le vocabulaire de l'âme au xiie siècle." *Revue des Sciences Philosophiques et Théologiques* 41 (1957): 209–32.

——. *Nature, Man, and Society in the Twelfth Century*. J. Taylor and L. Little, trans. Chicago: University of Chicago Press, 1968.

Chinca, Mark. *Meditating Death in Medieval and Early Modern Devotional Writing: From Bonaventure to Luther*. Oxford: Oxford University Press, 2020.

Clanchy, Michael T. *Looking Back from the Invention of Printing: Mothers and the Teaching of Reading in the Middle Ages*. Utrecht Studies in Medieval Literacy 40. Turnhout: Brepols, 2018.

Clark, Andy. *Being There: Putting Brain, Body, and World Together Again*. Cambridge, MA: MIT Press, 1997.

——. *Supersizing the Mind: Embodiment, Action, and Cognitive Extension*. Oxford: Oxford University Press, 2008.

——. *Mindware: An Introduction to the Philosophy of Cognitive Science*. Oxford: Oxford University Press, 2014.

Clark, Elizabeth A. *Reading Renunciation: Asceticism and Scripture in Early Christianity*. Princeton: Princeton University Press, 1999.

Clucas, Stephen. "Liquid History: Serres and Lucretius." In Niran Abbas, ed. *Mapping Michel Serres*, 72–83. Ann Arbor: University of Michigan Press, 2005.

Cocker, Emma. "Weaving Codes/Coding Weaves: Penelopean *Mêtis* and the Weaver-Coder's *Kairos*." *Textile* 15.2 (2017): 124–41.

Colish, Marcia L. *The Stoic Tradition from Antiquity to the Early Middle Ages*. 2 vols. Leiden: Brill, 1990.

——. "*Habitus* Revisited: A Reply to Cary Nederman." *Traditio* 48 (1993): 77–92.

Collins, Roger. *Visigothic Spain, 409- 711*. London: Blackwell, 2004.

Conybeare, Catherine. *The Irrational Augustine*. Oxford: Oxford University Press, 2006.

——. "*Quotquot haec legerint meminerint*: All who read this will remember." *MLN* 2012, no. 5 (Comparative Literature Issue Supplement): S23–S33.

——. "The Creation of Eve." *Augustinian Studies* 49.2 (2018): 181–98.

Copeland, Rita. *Rhetoric, Hermeneutics and Translation in the Middle Ages*. Cambridge: Cambridge University Press, 1991.

——. "The History of Rhetoric and the Longue Durée: Ciceronian Myth and Its Medieval Afterlives." *JEGP* 106.2 (2007): 176–202.

——. "*Pathos* and Pastoralism: Aristotle's *Rhetoric* in Medieval England." *Speculum* 89 (2014): 96–127.

—— and Ineke Sluiter eds. *Medieval Grammar and Rhetoric: Language Arts and Literary Theory, Ad 300–1475*. Oxford: Oxford University Press, 2009.

Corbett, Edward P. J. "The Rhetoric of the Open Hand and the Rhetoric of the Closed Fist." *College Composition and Communication* 20.5 (1969): 288–96.

Cornford, F. M. *Plato's Cosmology: The Timaeus of Plato*. London: Routledge, 2000 (orig publ. 1937).

Cornish, Alison. *Reading Dante's Stars*. New Haven: Yale University Press, 2000.

——. "*Paradiso* 29: Saving the Appearances." *Dante Studies* 137 (2020): 107–23.

Courcelle, Pierre. *Les Confessions de Saint Augustin dans la tradition littéraire*. Paris: Etudes Augustiennes, 1963.

——. *La Consolation de Philosophie dans la tradition littéraire*. Paris: Etudes Augustiennes, 1967.

Cox, Virginia and John O. Ward, eds. *The Rhetoric of Cicero in Its Medieval and Early Renaissance Commentary Tradition*. Leiden: Brill, 2006.

Crombie, A. C. (1953). *Robert Grosseteste and the Origins of Experimental Science 1100–1700*. Oxford: Oxford University Press.

Cross, Richard K. *Duns Scotus's Theory of Cognition*. Oxford: Oxford University Press, 2014.

Crouse, R. D. "Honorius Augustodunensis: The arts as *via ad patriam*." In *Arts libéraux et philosophie au moyen age*, 531–9. Congrès International de Philosophie Médiévale, 4. Montréal: Vrin, 1969.

Dan, Anca C. "La Mappemonde d'Albi – Un Pinax Chôrographikos." CFC-234 (Dec. 2017).

Darlington, Oscar G. "Gerbert, the Teacher". *The American Historical Review* 52 (1947): 456–76.

Del Puppo, Dario. "The Limits of Allegory in Jacobus de Cessolis' *De ludo scaccorum*". In D. E. O'Sullivan, *Chess in the Middle Ages and Early Modern Age: A Fundamental Thought Paradigm of the Premodern World*. Berlin: De Gruyter, 2012: 221–40.

Dinshaw, Carolyn. *How Soon Is Now? Medieval Texts, Amateur Readers, and the Queerness of Time*. Durham, NC: Duke University Press, 2012.

Diringer, David. *The Book before Printing*. New York: Dover Books, 1982.

Discenza, Nicole Guenther. "The Old English Boethius." In N. F. Discenza and P. Szarmach, eds. *A Companion to Alfred the Great*, 200–26. Leiden: Brill, 2015.

Draaisma, Douwe. *Metaphors of Memory: A History of Ideas about the Mind*. Trans. Paul Vincent. Cambridge: Cambridge University Press, 2000.

Drake, Stillman. *Galileo*. Oxford: Oxford University Press, 1980.

Dronke, Peter. *Fabula: Explorations into the Uses of Myth in Medieval Platonism*. Mittellateinische Studien und Texte 9. Leiden: Brill, 1974.

——. "The Symbolic Cities of Hildegard of Bingen." *The Journal of Medieval Latin* 1 (1991): 168–83.

——. *The Spell of Calcidius: Platonic Concepts and Images in the Medieval West*. Millennio Medievale 74. Firenze: SISMEL: Galluzzo, 2008.

Dudai, Yadin. *Memory from A to Z: Keywords, Concepts and Beyond*. Oxford: Oxford University Press, 2002.

—— and Mary Carruthers. "Concepts: The Janus Face of Mnemosyne." *Nature* 434 (March 31, 2005): 567.

Eden, Kathy. *Poetic and Legal Fiction in the Aristotelian Tradition*. Princeton: Princeton University Press, 1986.

Edson, Evelyn. *Mapping Time and Space: How Medieval Mapmakers Viewed Their World*. London: British Library, 1997.

Edwards, Robert R. *Invention and Authorship in Medieval England*. Columbus: Ohio State University Press, 2017.

Ekirch, A. Roger. *At Day's Close: Night in Times Past*. New York: Norton, 2005.

Esler, Philip F. "Pacino Di Bonaguida's Tree of Life: Interpreting the Bible in Paint in Early Fourteenth-Century Italy." *Biblical Reception* 3 (2015): 1–29. Online at http://eprints.glos.ac.uk/id/eprint/2857.

Evans, G. R. "The 'sub-Euclidean' Geometry of the Earlier Middle Ages up to the Mid-Twelfth Century." *Archive for the History of Exact Sciences* 16 (1976): 105–18.

——. "The Rithmomachia: A Mediaeval Mathematical Teaching Aid?" *Janus: Revue internationale de l'histoire des sciences, de la médecine, de la pharmacie, et de la technique* 63 (1976): 257–71.

——. "The *Saltus Gerberti*: The Problem of the 'Leap.'" *Janus: Revue internationale de l'histoire des sciences, de la médecine, de la pharmacie, et de la technique* 67.4 (1980): 261–8.

——. *Alan of Lille*. Cambridge: Cambridge University Press, 1983.

Evans, Michael W. "The Geometry of the Mind." *Architectural Association Quarterly* 12.4 (1980): 32–55.

Even-Ezra, Ayelet. *Lines of Thought: Branching Diagrams and the Medieval Mind*. Chicago: University of Chicago Press, 2021.

Faller, Stephan. "The World According to Cosmas Indicopleustes – Concepts and Illustrations of an Alexandrian Merchant and Monk." *Transcultural Studies* (2011): 193–32.

Fassler, Margot E. "Allegorical Architecture in *Scivias*: Hildegard's Setting for the *Ordo Virtutum*." *Journal of the American Musicological Society* 67.2 (2014): 317–78. https://doi.org/DOI: 10.1525/jams.2014.67.2.317.

——. *Cosmos, Liturgy, and the Arts in the Twelfth Century*. Philadelphia: University of Pennesylvania Press, 2022.

Flash, Tamar and Alain Berthoz, eds. *Space-Time Geometries for Motion and Perception in the Brain and the Arts*. Springer, 2021.

Flobert, Pierre. *Les verbes déponents latins des origines à Charlemagne*. Paris: Les Belles Lettres, 1975.

Folkerts, Menso. *The Development of Mathematics in Medieval Europe: The Arabs, Euclid, Regiomontanus*. Aldershot: Ashgate, 2006.

—— and A. J. E. M. Smeur. "Franco Leodiensis *De Quadratura Circuli*." *Archives internationales d'histoire des sciences* 26 (1976): 61–98.

Fontaine, Jacques. *Isidore de Seville et la culture classique dans l'Espagne Wisigothique*. 3 vols. Paris: Etudes Augustiniennes, 1983.

——. *Isidore de Séville: genèse et originalité de la culture hispanique au temps des Wisigoths*. Turnhout: Brepols, 2000.

Franklin, Carmela V. "Grammar and Exegesis: Bede's *Liber de schematibus et tropis*." In C. D. Lanham, ed. *Latin Grammar and Rhetoric: Classical Theory and Medieval Practice*, London: Continuum, 2002.

Franklin, James. "Diagrammatic Reasoning and Modelling in the Imagination: The Secret Weapons of the Scientific Revolution." In Guy Freeland and Anthony Corones, eds. *1543 and All That: Images and Word, Change and Continuity in the Proto-Scientific Revolution*, 53–115. Dordrecht and London: Kluwer Academic. 2000.

Fredborg, Karin M. "The Scholastic Teaching of Rhetoric in the Middle Ages." *CIMAGL* 55 (1978): 85–105.

Frede, Dorothea. "The Cognitive Role of *Phantasia* in Aristotle." In M. C. Nussbaum and A. O. Rorty, eds. *Essays on Aristotle's De Anima*, 279–95. Oxford: Oxford University Press, 1992.

——. "Mixed Feelings in Aristotle's Rhetoric." In A. O. Rorty, ed. *Essays on Aristotle's Rhetoric*, 258–85. Berkeley: University of California Press, 1996.

Friedman, John Block. "The Architect's Compass in Creation Miniatures of the Later Middle Ages." *Traditio* 30 (1974): 419–29.

Gaspar, Giles E. M. "Creation, Light, and Redemption: Hexaemeral Thinking, Robert Grosseteste, and the *Summa Halensis*." In L. Schumacher, ed. *The Summa Halensis: Sources and Context*, 295–316. Berlin: De Gruyter, 2020.

Gautier-Dalché, Patrick. "De la glose à la contemplation. Place et fonction de la carte dans les manuscrits du haut moyen âge." In *Testo e immagine nell'alto medioevo* (2 vols.) *Settimane di studio del centro italiano di studi sull' alto medioeva* 41: 693–771. Spoleto, 1994.

Gehl, Paul F. "'Competens Silentium': Varieties of Monastic Silence in the Medieval West." *Viator* 18 (1987): 125–60.

Gilson, Etienne. *The Christian Philosophy of St. Augustine*. New York: Random House, 1960.

Goldie, Matthew B. *Scribes of Space: Place in Middle English Literature and Late Medieval Science*. Ithaca: Cornell University Press, 2019.

Grattan-Guiness, Ivor. "Numbers, Magnitudes, Ratios, and Proportions in Euclid's Elements: How Did He Handle Them?" *Historia Mathematica*, 23 (1996): 355–75.

Graves, Michael. "Architecture and the Lost Art of Drawing." *New York Times Sunday Review*, September 2, 2012.

Grosfillier, Jean. "Quelques considérations sur l'influence du *De contemplatione* de Richard de Saint-Victor." *Sacris Erudiri* 52 (2013): 235–74.

Hadot, Pierre. *Philosophy as a Way of Life*. Oxford: Blackwell, 1995.

Hamburger, Jeffrey. "*Haec figura demonstrat': Diagramme in einem Pariser Exemplar von Lothars von Segni, "De Missarum Mysteriis' aus dem frühen 13. Jahrhundert*. Berlin: De Gruyter, 2013.

——. *Diagramming Devotion: Berthold of Nuremberg's Transformation of Hrabanus Maurus's Poems in Praise of the Cross*. Chicago: University of Chicago Press, 2020.

Hamburger, Jeffrey, L. Safran, and D. Roxburgh, eds. *The Diagram as Paradigm*. Washington, DC: Dumbarton Oaks Research Library and Collection, 2022.

Harlizius-Klück, Ellen. *Weberei als Episteme und die Genese der Deduktiven Mathematik*. Berlin: Edition Eberbach, 2004.

——. "Arithmetics and Weaving: From Penelope's Loom to Computing." *Münchner Wissenschaftstage*, no. 8 (October 18–21, 2008): 6 pp.

Harries, Karsten. "The Infinite Sphere: Comments on the History of a Metaphor." *Journal of the History of Philosophy* 13, no. 1 (1975): 5–15.

Hawhee, Debra. "Looking into Aristotle's Eyes: Toward a Theory of Rhetorical Vision." *Advances in the History of Rhetoric* 14.2 (2011): 139–65. https://doi.org/10.1080/15362426.2011.613288.

Heller, Agnes. "The Metaphor of the Throw in Nicholas of Cusa's *Game of Spheres*," *Graduate Faculty Philosophy Journal* 33.2 (2012): 473–90.

Heyes, Cecilia M. *Cognitive Gadgets: The Cultural Evolution of Thinking*. Cambridge, MA: Harvard University Press, 2018.

Hicks, Andrew J. *Composing the World: Harmony in the Medieval Platonic Cosmos*. Oxford: Oxford University Press, 2017.

Hollingworth, Miles. *Saint Augustine of Hippo: An Intellectual Biography*. Oxford: Oxford University Press, 2013.

Hunt, Richard W. *The History of Grammar in the Middle Ages: Collected Papers*, ed. G. L. Bursill-Hall. Amsterdam: Benjamins, 1980.

—— and R. Klibansky. "An 'Autobiography' of Guido Faba." *JWCI* 1 (1941–43): 253–80.

Iribarren, Isabel. "The Cult of the Marriage of Joseph and Mary: The Shaping of Doctrinal Novelty in Jean Gerson's *Josephina* (1414–17)." In A. Fitzpatrick and J. Sabapathy, eds. *Individuals and Institutions in Medieval Scholasticism*, 253–68. London: IHR, University of London Press, 2020.

Jacob, Christian. *L'empire des cartes: Approche théorique de la cartographie à travers l'histoire*. Paris: Albin Michel, 1992.

Jacquart, Danielle. "La notion philosophico-médicale de *spiritus* dans l'Avicenne latin." In G. Gubbini, ed. *Body and Spirit in the Middle Ages*, 13–33. Berlin: De Gruyter, 2020.

Jaeger, C. Stephen. *The Envy of Angels: Cathedral Schools and Social Ideals in Medieval Europe, 950–1200*. Philadelphia: University of Pennsylvania Press, 1994.

Jager, Eric. *The Book of the Heart*. Chicago: University of Chicago Press, 2000.

Jeauneau, Edouard. *Rethinking the School of Chartres*. Toronto: University of Toronto Press, 2020. https://doi.org 10.3138/9781442606746.

Johnson, Eleanor. *Staging Contemplation: Participatory Theology in Middle English Prose, Verse, and Drama*. Chicago: University of Chicago Press, 2018.

Johnston, Michael and M. Van Dussen, eds. *The Medieval Manuscript Book: Cultural Approaches*. Cambridge: Cambridge University Press, 2015.

Jones, Peter Murray. "*Complexio* and *Experimentum*: Tensions in Late Medieval Medical Practice." In P. Horden and E. Hsu, eds. *The Body in Balance: Humoral Medicines in Practice*, 107–28. Oxford: Berghahn, 2013.

Joyner, Danielle. *Painting the Hortus Deliciarum: Medieval Women, Wisdom, and Time*. University Park: Pennsylvania State University Press, 2016.

Jung, Jacqueline E. "The Tactile and the Visionary: Notes on the Place of Sculpture in the Medieval Religious Imagination." In C. Hourihane, ed. *Looking Beyond: Visions, Dreams, and Insights in Medieval Art and History*, 202–40. Princeton: Index of Christian Art, 2010.

——. *Eloquent Bodies: Movement, Expression, and the Human Figure in Gothic Sculpture*. New Haven and London: Yale University Press, 2020.

Kaster, Robert A. *Guardians of Language: The Grammarian and Society in Late Antiquity*. Berkeley and London: University of California Press, 1988.

Katz, Victor J., Menso Folkerts et al., eds. *Sourcebook in the Mathematics of Medieval Europe and North Africa*. Princeton: Princeton University Press, 2016.

Kay, Sarah. *The Place of Thought: The Complexity of One in Late Medieval French Didactic Poetry*. Philadelphia: University of Pennsylvania Press, 2007.

Kaye, Joel. *A History of Balance 1250–1375*. Cambridge: Cambridge University Press, 2014.

Kelly, Douglas. "Topical Invention in Medieval French Literature." In J. J. Murphy, ed. *Medieval Eloquence*, 231–51. Berkeley: University of California Press, 1978.

——. "The Scope of Medieval Instruction in the Art of Poetry and Prose: Recent Developments in Documentation and Interpretation." *Studies in Medieval and Renaissance Teaching; SMART* n. s. 6 (1998): 49–68.

Kelly, Louis G. *The Mirror of Grammar: Theology, Philosophy and the Modistae*. Amsterdam: Benjamins, 2002.

Kelly, Peter. "Crafting *Chaos*: Intelligent Design in Ovid, *Metamorphoses* Book I and Plato's *Timaeus*." *The Classical Quarterly* 70.2 (2021): 734–48.

Kempshall, Matthew S. "The Virtues of Rhetoric: Alcuin's *Disputatio De rhetorica et De uirtutibus*." *Anglo-Saxon England* 37 (2008): 7–30.

Kendrick, Laura. *Animating the Letter: The Figurative Embodiment of Writing from Late Antiquity to the Renaissance*. Columbus: Ohio State University Press, 1999.

Kennedy, George A. *A New History of Classical Rhetoric*. Princeton: Princeton University Press, 1994.

——. *Classical Rhetoric and Its Christian and Secular Tradition*. Second, revised and enlarged ed. Chapel Hill: University of North Carolina Press, 1999.

——. *Progymnasmata: Greek Textbooks of Prose Composition and rhe£toric*. Leiden: Brill, 2003.

Kenny, Anthony J. P. *Aquinas on Mind*. London: Routledge, 1993.

Kessler, Herbert A. *Spiritual Seeing: Picturing God's Invisibility in Medieval Art*. Philadelphia: University of Pennsylvania Press, 2000.

——. *Seeing Medieval Art*. Peterborough, ONT: Broadview Press, 2004.

——. "*Medietas/ Mediator* and the Geometry of Incarnation." In W. S. Melion and L. P. Wandel, eds. *Image and Incarnation: The Early Modern Doctrine of the Pictorial Image*, 17–75. Leiden: Brill, 2015.

Kidson, Peter. "A Metrological Investigation." *JWCI* 53 (1990): 71–97.

——. "Architectural Proportion: Before 1450." *Grove Art Online* (2003): 1–15.

Kinsella, Karl. "Richard of St Victor's Solutions to Problems of Architectural Representation in the Twelfth Century." *Architectural History* 59 (2016): 3–24.

Kline, Naomi Reed. *Maps of Medieval Thought: The Hereford Paradigm*. Suffolk: Boydell, 2001.

Knorr, Wilbur. "The Geometry of Burning-Mirrors in Antiquity." *Isis* 74.1 (1983): 53–73.

Kominko, Maja. *The World of Kosmas: Illustrated Byzantine Codices of the Christian Topography*. Cambridge: Cambridge University Press, 2013.

Kotschick, Dieter. "The Topology and Combinatorics of Soccer Balls." *American Scientist* 94, no. 4 (2006): 350–7. https://doi.org/10.1511/2006.60.350.

Krämer, Sybille. "'The Mind's Eye': Visualizing the Non-Visual and the 'Epistemology of the Line.'" In Richard Heinrich, Elisabeth Nemeth et al., eds. *Image and Imaging in Philosophy, Science and the Arts*, 275–93. Frankfurt: ontos Verlag, 2011.

Kühnel, Bianca. *The End of Time in the Order of Things*. Regensburg: Schnell & Steiner, 2003.

Kupfer, Marcia. *Art and Optics in the Hereford Map*. New Haven: Yale University Press, 2016.

——. "The Cosmic Vision of Saint Benedict, *e specula* and *in speculo*." In N. Bouloux et al., eds. *Orbis Disciplinae. Liber amicorum Patrick Gautier Dalché*, 139–65. Turnhout: Brepols, 2017.

——. "Traveling the Mappa Mundi: Readerly Transport from Cassiodorus to Petrarch." In N. Debby et al., eds. *Maps and Travel in the Middle Ages and the Early Modern Period*, 17–36. Berlin: De Gruyter, 2018.

——. A. S. Cohen and J. H. Chajes, eds. *The Visualization of Knowledge in Medieval and Early Modern Europe*. Turnhout: Brepols, 2020.

Lakey, Christopher. *Sculptural Seeing: Relief, Optics, and the Rise of Perspective in Medieval Italy*. New Haven: Yale University Press, 2018.

Lakoff, George. *Women, Fire, and Dangerous Things*. Chicago: University of Chicago Press, 1987.

Lakoff, George and Mark Johnson. *Metaphors We Live By*. Chicago: University of Chicago, 1980.

Lanham, Carol D. "Freshman Composition in the Early Middle Ages: Epistolography and Rhetoric before the *Ars Dictaminis*." *Viator* 23 (1992): 115–34.

——. "Writing Instruction from Late Antiquity to the Twelfth Century." In J. J. Murphy, ed. *A Short History of Writing Instruction*, 79–121. Mahwah, NJ: Erlbaum, 2001.

——, ed. *Latin Grammar and Rhetoric: From Classical Theory to Medieval Practice*. London and New York: Continuum, 2002.

Lapidge, Michael. "A Stoic Metaphor in Late Latin Poetry: The Binding of the Cosmos." *Latomus* 39.4 (1980): 817–37.

Lausberg, Heinrich. *A Handbook of Literary Rhetoric*. Trans. M. T. Bliss et al. Leiden: Brill, 1998.

Law, Vivien. *Grammar and Grammarians in the Early Middle Ages*. London: Longman, 1997.

Lawrence, C. H. *St. Edmund of Abingdon: A Study in Hagiography and History.* Oxford: Clarendon Press, 1960.

Leach, Elizabeth Eva. *Sung Birds: Music, Nature, and Poetry in the Later Middle Ages.* Ithaca: Cornell University Press, 2007.

——. "Nature's Forge and Mechanical Production: Writing, Reading and Performing Song." In Mary Carruthers, ed. *Rhetoric Beyond Words*, 72–95. Cambridge: Cambridge University Press, 2010.

Leclercq, Jean. *La Spiritualité de Pierre de Celle.* Paris: Vrin, 1946.

——. *Otia Monastica: études sur le vocabulaire de la contemplation au moyen age.* Rome: Herder, 1963.

——. *The Love of Learning and the Desire for God.* C. Misrahi, trans. 3rd ed. New York: Fordham University Press, 1982.

LeDoux, Joseph. *The Emotional Brain.* New York: Simon and Schuster, 1996.

——. "Rethinking the Emotional Brain." *Neuron* 73 (February 23, 2012): 653–76.

Leff, Michael C. "The Topics of Argumentative Invention in Latin Rhetorical Theory from Cicero to Boethius." *Rhetorica* 1.1 (1983): 23–44.

Le Goff, Jacques. *Time, Work & Culture in the Middle Ages.* A. Goldhammer, trans. Chicago: University of Chicago Press, 1980.

Lesieur, Thierry. "Raison et rationalité chez Hugues de Saint-Victor." In D. Poirel et al., eds. *L'école de Saint-Victor de Paris: Influence et rayonnement du moyen âge á l'époque moderne,* 385–403. Bibliotheca Victorina. Turnhout: Brepols, 2010.

Leyser, Conrad. "*Lectio Divina, Oratio Pura*: Rhetoric and the Techniques of Asceticism in the *Conferences* of John Cassian." In M. C. Barone and F. S. Barcellona, eds. *Modelli di Santità e Modelli di Comportamento: Contrasti, Intersezioni, Complementarità*, 79–105. Turin: Rosenberg & Sellier, 1994.

Lidov, Alexei. "The Holy Fire and Visual Constructs of Jerusalem, East and West." In B. Kuehnel et al., eds. *Visual Constructs of Jerusalem*, 241–9. Turnhout: Brepols, 2014.

Lindberg, David. *Theories of Vision from Al-Kindi to Kepler.* Chicago: University of Chicago Press, 1976.

Lombardi, Elena. *The Syntax of Desire: Language and Love in Augustine, the Modistae, Dante.* Toronto: University of Toronto Press, 2007.

Louth, Andrew. *The Origins of the Christian Mystical Tradition from Plato to Denys.* Oxford: Oxford University Press, 1981.

Lugli, Emanuele. *The Making of Measure and the Promise of Sameness.* Chicago: University of Chicago Press, 1996.

Lüthy, Christoph and Alexis Smets. "Words, Lines, Diagrams, Images: Towards a History of Scientific Imagery." *Early Science and Medicine* 14 (2009): 398–439.

Lutz, Eckhart C., Vera Jerjen, and Christine Putzo, eds. *Diagramm und Text.* Wiesbaden: Reichert Verlag, 2014.

MacCulloch, Diarmaid. *A History of Christianity: The First Three Thousand Years*. London: Allen Lane, 2009.

Mack, Peter. *Renaissance Argument: Valla and Agricola in the Traditions of Rhetoric and Dialectic*. Leiden: Brill, 1993.

——. *A History of Renaissance Rhetoric, 1380–1620*. Oxford: Oxford University Press, 2011.

Marenbon, John. *Later Medieval Philosophy*. London: Routledge, 1987.

——. *Boethius*. Oxford: Oxford University Press, 2003.

Margolis, Howard. *Patterns, Thinking, and Cognition*. Chicago: University of Chicago, 1987.

Marrou, Henri. *A History of Education in Antiquity*. George Lamb, trans. Madison, WI: University of Wisconsin Press, 1982.

——. *Saint Augustin et la fin de la culture antique*. 4th ed. Paris: Boccard, 1958.

Martinez, Ronald L. "Dante 'Buon Sartore' (*Paradiso* 32.140): Textile Arts, Rhetoric, and Metapoetics at the End of the *Commedia*." *Dante Studies* 136 (2018): 22–61.

Masi, Michael. "A Newberry Diagram of the Liberal Arts." *Gesta* 11 (1972): 52–6.

Mayr-Harting, Henry. "Ruotger, the Life of Bruno, and Cologne Cathedral Library.'" In L. J. Smith and B. Ward, eds. *Intellectual Life in the Middle Ages: Essays Presented to Margaret Gibson*, 33–60. London, 1992.

——. *Church and Cosmos in Early Ottonian Germany: The View from Cologne*. Oxford: Oxford University Press, 2007.

McGinn, Bernard. *The Golden Chain: A Study in the Theological Anthropology of Isaac of Stella*. Washington: Cistercian Publications, 1972.

——, ed. and trans. *Three Treatises on Man: A Cistercian Anthropology*. Kalamazoo, MI: Cistercian Publications, 1977.

——. "Does the Trinity Add Up? Transcendental Mathematics and Trinitarian Speculation." In E. R. Elder, ed. *Praise No Less Than Charity: Studies in Honor of M. Chrysogonus Waddell*, 237–64. Kalamazoo: Cistercian Publications, 2002.

——. *The Presence of God: A History of Western Christian Mysticism*. 4 vols. New York: Crossroad, 1991–2005. Vol. 1, *The Foundations of Mysticism*, 1991; vol. 2, *The Growth of Mysticism: Gregory the Great through the 12th Century*, 1994; vol. 3, *The Flowering of Mysticism*, 1998; vol. 4, *The Harvest of Mysticism in Medieval Germany (1300–1500)*, 2005.

McGuire, Brian Patrick. *Friendship and Community: The Monastic Experience, 350–1250*. Kalamazoo, MI: Cistercian Publications, 1988.

McKeon, Richard. "Arts of Invention and Arts of Memory: Creation and Criticism." *Critical Inquiry* 1 (1975): 723–39.

McKitterick, Rosamond, ed. *The Uses of Literacy in Early Medieval Europe*. Cambridge: Cambridge University Press, 1990.

McLeish, Tom, H. Smithson et al. "A Thirteenth-Century Theory of Speech." *Journal of the Acoustical Society of America* 146.2 (2019): 237–47.

—— and Mary Garrison. "Reversals in Wartime: Alcuin and Charlemagne Discuss Retrograde Motion." *Interfaces* 8.3 (2021): 14–53.

—— with Tom Lancaster and Mark Pexton. "Emergence and Topological Order in Classical and Quantum Systems." *Studies in the History and Philosophy of Modern Physics* 66 (2019): 155–69.

McNamee, Megan C. "Picturing as Practice: Placing a Square above a Square in the Central Middle Ages." In A. and G. W. Most, eds. *Canonical Texts and Scholarly Practices: A Global Comparative Approach*, 200–23. Cambridge: Cambridge University Press, 2016.

——. "Imaging and Imagining Solidity." In B. Kitzinger and J. O'Driscoll, eds. *After the Carolingians: Re-Defining Manuscript Illumination in the 10th and 11th Centuries*, 86–117. Berlin: DeGruyter, 2019.

——. "Early Romanesque Abstraction and the 'Unconditionally Two-Dimensional Surface.'" In E. Gertsman, ed. *Abstraction in Medieval Art: Beyond the Ornament*, 267–84. Amsterdam: Amsterdam University, 2020.

Meier, Christel. "*Respice Principium*: Die Bedeutung des Anfangs im prophetischen Werk Hildegards von Bingen." *Frühmittelalterliche Studien* 47 (2013): 185–208.

Mews, Constant, "The Diffusion of the *De spiritu et anima* and Cistercian reflection on the soul," *Viator* 49.3 (2018): 297–330.

Minnis, Alastair. *Medieval Theory of Authorship.* 2nd ed. Philadelphia: University of Pennsylvania Press, 1988.

Mulchahey, Michèle M. *"First the Bow Is Bent in Study ...": Dominican Education before 1350.* Studies and Texts, 132. Toronto: Pontifical Institute of Mediaeval Studies, 1998.

Murphy, James J. *Rhetoric in the Middle Ages.* Berkeley: University of California Press, 1974.

——, ed. *A Short History of Writing Instruction* (2nd ed.). Mahwah, NJ: Erlbaum, 2001.

Murray, Alexander. *Reason and Society in the Middle Ages.* Oxford: Oxford University Press, 1978.

Murray, Stephen. *Notre Dame, Cathedral of Amiens: The Power of Change in Gothic.* Cambridge: Cambridge University Press, 1996.

——. *Plotting Gothic.* Chicago: University of Chicago Press, 2014. doi: 10.7208/chicago/9780226191942.001.0001.

Nash, Ronald H. *The Light of the Mind: St. Augustine's Theory of Knowledge.* Lexington, KY: University Press of Kentucky, 1969.

Nederman, Cary. "Nature, Ethics, and the Doctrine of '*Habitu*': Aristotelian Moral Psychology in the Twelfth Century." *Traditio* 45 (1989–90): 87–110.

Netz, Reviel. *The Shaping of Deduction in Greek Mathematics.* Cambridge: Cambridge University Press, 1999.

——. "Imagined and Layered Ontology in Greek Mathematics." *Configurations* 17 (2009a): 19–50.

——. *Ludic Proof: Greek Mathematics and the Alexandrian Aesthetic*. Cambridge: Cambridge University Press, 2009b.

Netz, Reviel and William Noel. *The Archimedes Codex*. London: Da Capo Press, 2007.

Newman, Barbara. "What Did It Mean to Say 'I Saw'? The Clash between Theory and Practice in Medieval Visionary Culture." *Speculum* 80 (2005): 1–43.

——. "Contemplating the Trinity: Text, Image, and the Origins of the Rothschild Canticles". *Gesta* 52 (2013): 133–59.

Newman, Sara. "Aristotle's Notion of 'Bringing before the Eyes': Its Contributions to Aristotelian and Contemporary Conceptualizations of Metaphor, Style, and Audience." *Rhetorica* 20 (2002): 1–23.

Novikoff, Alex J. *The Medieval Culture of Disputation: Pedagogy, Practice, and Performance*. Philadelphia: University of Pennsylvania, 2013.

Nussbaum, M. C. and A. O. Rorty, eds. *Essays on Aristotle's De Anima*. Oxford: Clarendon Press, 1992.

Oberman, Heiko A. *The Harvest of Medieval Theology: Gabriel Biel and Late Medieval Theology,* revised ed. Cambridge: Harvard University Press, 1965.

Obrist, Barbara. "Le diagramme isidorien de l'année et des saisons: Son contenu physique et les représentations figuratives." *Mélanges de l'École française de Rome*: *Moyen âge* 108.1 (1996): 95–164.

——. "Wind Diagrams and Medieval Cosmology." *Speculum* 72, no. 1 (1997): 33–84.

——. *La cosmologie médiévale, textes et images: Les fondements antiques*. Firenze: SISMEL, Edizioni del Galluzzo, 2004.

——. "Graphic Vehicles of *Scientia*: The Idea of a Spherical Universe and Its Visualization in the Earlier Middle Ages (Seventh–Twelfth Centuries)." In M. Kupfer, A. Cohen et al., eds. *The Visualization of Knowledge* (q.v.), 229–58.

O'Daly, Irene. "Diagrams of Knowledge and Rhetoric in Manuscripts of Cicero's *De Inventione*." In E. Kwakkel, ed. *Manuscripts of the Latin Classics 800–1200*, 77–105. Leiden: Leiden University Press, 2015.

Ohly, Friedrich. *Geometria e Memoria: Lettera e allegoria nel medioevo*. Translated by L. R. Santini. Bologna: Il Molino, 1985.

Ong, Walter J. *Ramus: Method and the Decay of Dialogue.* Cambridge, MA: Harvard University Press, 1958.

Oosterhoff, R. J. "*Idiotae*, Mathematics, and Artisans: The Untutored Mind and the Discovery of Nature in the Fabrist Circle." *Intellectual History Review*, Special Issue: "The Nature of Invention" (2014): 1–19.

Ortuzar Escudero, María José. "The Role of the *spiritus animalis* in Accounts of Perception: Medical and Philosophical Remarks from Three 13th Century Encyclopedias." *Vincent of Beauvais Newsletter*, no. 42 (2018): 4–16.

Palazzo, Eric. *Le souffle de Dieu: L'energie de la liturgie et l'art au moyen âge*. Paris: Cerf, 2020.

——. "Le Christ énergétique, la spirale et la monstrance." In R. Dekoninck, M-C. Claes, and B. Baert, eds. *Ornamenta Sacra*, 39–56. Leuven: Peeters, 2022.

——. *Broder la splendeur: La théologie chrétienne de l'ornement dans l'Antiquité et le haut Moyen Âge*. Paris: Cerf, 2024.

Pelikan, Jaroslav. *Christianity and Classical Culture: The Metamorphosis of Natural Theology in the Christian Encounter with Hellenism*. New Haven: Yale University Press, 1993.

——. *The Mystery of Continuity: Time and History, Memory and Eternity in the Thought of Saint Augustine*. Charlottesville: University Press of Virginia, 1986.

Penrose, Roger. *Cycles of Time: An Extraordinary New View of the Universe*. London: Bodley Head, 2010.

Pentcheva, Bissera. "Performative Images and Cosmic Sound in the *Exultet* Liturgy of Southern Italy." *Speculum* 95.3 (2020): 396–466.

Pollmann, Karla. "Christianity and Authority in Late Antiquity: The Transformation of the Concept of *Auctoritas*." In C. Humfress et al., eds. *Being Christian in Late Antiquity*, 156–74. Oxford: Oxford University Press, 2014.

Rabbow, Paul. *Seelenführung: Methodik Der Exerzitien in Der Antike*. Munich: Kösel Verlag, 1954.

Ramirez-Weaver, Eric. "William of Conches, Philosophical Continuous Narration, and the Limited Worlds of Medieval Diagrams." *Studies in Iconography* 30 (2009): 1–41.

Rampling, Jennifer M. *The Experimental Fire: Inventing English Alchemy, 1300–1700*. Chicago: University of Chicago Press, 2020.

Rankin, Susan K. *Writing Sounds in Carolingian Europe: The Invention of Musical Notation*. Cambridge: Cambridge University Press, 2018.

Rapp, Claudia. "Desert, City and Countryside in the Early Christian Imagination". *Church History and Religious Culture* 86 (2006): 93–112.

——. *Brother-making in Late Antiquity and Byzantium: Monks, Laymen, and Christian Ritual*. New York: Oxford University Press, 2016.

Rector, Geoffrey. "'En sa chambre souvent le lit': Literary Leisure and the Chamber Sociabilities of Early Anglo-French Literature (c. 1100–1150)." *Medium Aevum* 81 (2012): 88–125.

Reid, Pauline. "Through a Looking-Glass: Invention and Imagination in the Visual Rhetoric of William Caxton's *Mirrour and Description of the World* and Stephen Hawes's *Pastime of Pleasure*." *Rhetorica* 34.3 (2016): 268–300.

Reinhardt, Tobias. *Cicero's Topica*. Oxford Classical Monographs. Oxford: Oxford University Press, 2003.

Renardy, Charles. "Les ecoles liègoises du ixe au xiie siecle: grandes lignes de leur évolution." *Revue Belge de philologie et d'histoire* 57 (1979): 309–28.

Reydams-Schils, Gretchen. *Calcidius on Plato's Timaeus: Greek Philosophy, Latin Reception, and Christian Contexts.* Cambridge: Cambridge University Press, 2020.

Reynolds, Suzanne. *Medieval Reading: Grammar, Rhetoric, and the Classical Text.* Cambridge: Cambridge University Press, 1996.

Riché, Pierre. *Education and Culture in the Barbarian West: From the Sixth through the Eighth Century.* J. Contreni, trans. Columbia, SC: University of South Carolina Press, 1978.

——. *Gerbert d'Aurillac, Le Pape de l'an mil.* Paris: Fayard, 1987.

——. *Écoles et enseignement dans le haut moyen age: fin du ve siècle–milieu du xie siècle.* Paris: Picard, 1989.

Ricoeur, Paul. *Time and Narrative.* K. McLaughlin and D. Pellauer, trans. Chicago: University of Chicago Press, 1984.

Roediger, Henry L., Yadin Dudai, and Susan M. Fitzpatrick, eds. *Science of Memory: Concepts.* Oxford Scholarship Online: Oxford University Press, 2009.

Rosenwein, Barbara H. *Emotional Communities in the Early Middle Ages.* Ithaca: Cornell University Press, 2006.

——. "Emotion Words." In P. Nagy and D. Boquet, eds. *Le sujet des émotions au Moyen Âge*, 93–106. Paris: Beauchesne, 2008.

Rosier-Catach, Irène. *La grammaire spéculative des Modistes.* Lille: Presses universitaires de Lille, 1983.

——. *La parole comme acte: sur la grammaire et la sémantique au xiiie siècle.* Paris: Vrin, 1994.

——. "Discussions médiévales sur l'expression des affects." In P. Nagy and D. Boquet, eds. *Le sujet des émotions au moyen âge*, 201–23. Paris: Beauchesne, 2008.

Rotman, Brian. *Mathematics as Sign: Writing, Imagining, Counting.* Stanford: Stanford University Press, 2000.

Rouse, Richard A. and Mary A. Rouse. "The Vocabulary of Wax Tablets." *Harvard Library Bullein* n.s. 1.3 (1990): 12–19.

Rovelli, Carlo. *The Order of Time.* S. Carnell and E. Segre, trans. London: Penguin (Allen Lane), 2018.

Rudolph, Conrad. "Theories and Images of Creation in the Twelfth Century." *Art History* 22.1 (1999): 3–55.

——. *The Mystic Ark: Hugh of Saint Victor, Art, and Thought in the Twelfth Century.* Cambridge: Cambridge University Press, 2014.

Saiber, Arielle and Aba Mibrika. "The Three *Giri* of *Paradiso* 33." *Dante Studies* 131 (2013): 237–72 + 28 figures.

Sanchez Ameijeiras, Rocio. "Rebellious Architecture: Movement, Subversion and Transubstantiation in the Visionary Mode." *Codex Aquilarensis* 35 (2019): 85–113.

Sandler, Lucy Freeman. "Religious Instruction and Devotional Study: The Pictorial and the Textual in Gothic Diagrams." In M. Kupfer et al., eds. (q.v.), *The Visualization of Knowledge in Medieval and Early Modern Europe*, 429–48.

Sanford, Eva M. "De Loquela Digitorum." *The Classical Journal* 23.8 (1928): 588–93.

Scarry, Elaine. "The Well-Rounded Sphere: The Metaphysical Structure of the *Consolation of Philosophy*." In C. D. Eckhardt, ed., *Essays in the Numerical Criticism of Medieval Literature*, 91–140. Lewisburg, PA: Bucknell University Press, 1980.

Scholem, Gershon. *Major Trends in Jewish Mysticism*. New York: Schoken Books, 1974. 1941.

Schulz, Verena. "Rhetoric and Medicine – The Voice of the Orator in Two Ancient Discourses." *Rhetorica* 34.2 (2016): 141–62.

Sears, Elizabeth. "The Iconography of Auditory Perception in the Early Middle Ages: On Psalm Illustration." In M. Fend, C. Burnett, and P. Gouk, eds., *The Second Sense: Studies in Hearing and Musical Judgement from Antiquity to the Seventeenth Century*, 19–42 (figs. 1–8). London: Warburg Institute, 1991.

Serres, Michel. *Hermes – Literature, Science, Philosophy*. Baltimore: Johns Hopkins University Press, 1982.

Shailor, Barbara A. and John Williams. *A Spanish Apocalypse: The Morgan Beatus Manuscript*. New York: Brazziler, 1991.

Shelby, Lon. "The Geometrical Knowledge of Mediaeval Master Masons." *Speculum* 47.3 (1972): 395–412.

Sherman, Claire Richter and Peter M. Lukehart, eds. *Writing on Hands: Memory and Knowledge in Early Modern Europe*. Washington, D.C.: Folger Shakespeare Library, 2000.

Sicard, Patrice. *Diagrammes médiévaux et exégèse visuelle: le "Libellus de formatione arche" de Hugues de Saint-Victor*. Turnhout: Brepols, 1993.

Simpson, James. "From Reason to Affective Knowledge: Modes of Thought and Poetic Form in *Piers Plowman*." *Medium Aevum* 55.1 (1986): 1–23.

Smith, Gregory A. "How Thin is a Demon?" *Journal of Early Christian Studies* 16.4 (2008): 479–512.

Smith, Lesley. *Fragments of a World: William of Auvergne and His Medieval Life*. Chicago: University of Chicago Press, 2023.

—— et al. *Codices Boethiani: A Conspectus of Manuscripts of the Works of Boethius*, vol. 2. London: Warburg Institute, 2002.

Smithson, H. E, G. E. M. Gasper, and T. C. B. McLeish. "All the Colours of the Rainbow." *Nature Physics* 10 (2014): 539–41.

Sorabji, Richard. *Time, Creation, and the Continuum: Theories in Antiquity and in the Early Middle Ages*. Ithaca: Cornell University Press, 1983.

——. *Emotion and Peace of Mind*. Oxford: Oxford University Press, 2000.

Southerden, Francesca. "The Art of Rambling: Errant Thoughts and Entangled Passions in Petrarch's 'The Ascent of Mt Ventoux.'" In J. Morton, P. Knox, and D. Reeve, eds. *Medieval Thought Experiments*, 197–221. Leiden: Brill, 2018.

Southern, Richard W. *Saint Anselm: A Portrait in a Landscape*. Cambridge: Cambridge University Press, 1990.

Spanos, Margaret. "The Sestina: An Exploration of the Dynamics of Poetic Structure." *Speculum* 53.3 (1978): 545–57.

Sparavigna, Amelia Carolina. "A Discussion of a Geometric Shape That Became a Symbol Known as Mandorla or *Vesica Piscis*, Starting from a Pythagorean Point of View." *Zenodo* (2020): 1–29. https://doi.org/DOI: 10.5281/zenodo.4298852.

Staley, Lynn. "The Trope of the Scribe and the Question of Literary Authority in the Works of Julian of Norwich and Margery Kempe." *Speculum* 66.4 (1991): 820–38.

Stern, E. M. "Ancient Glass in a Philological Context." *Mnemosyne* 60 (2007): 341–406.

Stevens, Wesley. "Euclidean Geometry in the Early Middle Ages." In Marie-Thérèse Zenner, ed. *Villard's Legacy: Studies in Medieval Technology, Science, and Art in Memory of Jean Gimpel*, 226–64. Aldershot: Ashgate, 2004.

Stewart, Columba, OSB. "Imageless Prayer and the Theological Vision of Evagrius Ponticus." *Journal of Early Christian Studies* 9.2 (2001): 173–204.

——. "Evagrius Ponticus and the Eastern Monastic Tradition on the Intellect and the Passions." *Modern Theology* 27.2 (2011): 263–75.

Stock, Brian. *Myth and Science in the Twelfth Century.* Princeton: Princeton University Press, 1972.

——. *After Augustine: The Meditative Reader and the Text*. Philadelphia: University of Pennsylvania Press, 2001.

——. *The Integrated Self: Augustine, the Bible, and Ancient Thought*. Philadelphia: University of Pennsylvania Press, 2017.

Stone, Mark H. "The Cubit: A History and Measurement Commentary." *Journal of Anthropology* 2014: 11 pp. https://doi.org/http://dx.doi.org/10.1155/2014/489757.

Stump, Eleonore. *Dialectic and Its Place in the Development of Medieval Logic.* Ithaca: Cornell University Press, 1989.

Suddendorf, Thomas and Michael C. Corballis. "The Evolution of Foresight: What Is Mental Time Travel and Is It Unique to Humans?" *Behavioral and Brain Sciences* 30 (2007): 299–351.

Sutton, John. *Philosophy and Memory Traces.* Cambridge: Cambridge University Press, 1998.

——. "Exograms and Interdisciplinarity: History, the Extended Mind, and the Civilizing Process." In Richard Menary, ed. *The Extended Mind.* Cambridge, 189–225. MA: MIT Press, 2010.

Swartz, Michael D. *Mystical Prayer in Ancient Judaism: An Analysis of Ma⊠aseh Merkavah*. Tuebingen: Mohr, 1992.

Swift, Helen J. *Gender, Writing, and Performance: Men Defending Women in Late Medieval France, 1440–1538*. Oxford: Clarendon Press, 2008.

Szarmach, Paul. "Augustine's *Soliloquia* in Old English." In N. G. Discenza and P. Szarmach, eds. *A Companion to Alfred the Great*, 227–55. Leiden: Brill, 2015.

Tachau, Katherine H. *Vision and Certitude in the Age of Ockham: Optics, Epistemology, and the Foundations of Semantics, 1250–1345*. Leiden: Brill, 1988.

——. "God's Compass and Vana Curiositas: Scientific Study in the Old French *Bible Moralisée*." *Art Bulletin* 80.1 (1998): 7–33.

——. "Approaching Medieval Scholars' Treatment of Cognition". In M. C. Pacheco and J. Meirinhos, eds. *Intellect and Imagination in Medieval Philosophy*. Turnhout: Brepols, 2006: vol. 1, pp. 1–34.

Tanner, Brian K., R. G. Bower, and T. McLeish, G. Gasper. "Unity and Symmetry in the *De Luce* of Robert Grosseteste." In J. P. Cunningham and M. Hocknull, eds. *Robert Grosseteste and the Pursuit of Religious and Scientific Learning in the Middle Ages*, 3–20. Switzerland: Springer International, 2016.

Teeuwen, Mariken and Sinéad O'sullivan, eds. *Carolingian Scholarship and Martianus Capella: Ninth-Century Commentary Traditions on "De Nuptiis" in Context*. Turnhout: Brepols, 2012.

Truitt, Elly R. "Celestial Divination and Arabic Science in Twelfth-Century England: The History of Gerbert of Aurillac's Talking Head." *Journal of the History of Ideas* 73.2 (2012): 201–22.

——. *Medieval Robots: Mechanism, Magic, Nature, and Art*. Philadelphia: University of Pennsylvania Press, 2015.

——. "Knowledge and Power: Courtly Science and Political Utility in the Work of Roger Bacon." *Revista Española de Filosophíia Medieval* 28.1 (2021): 99–123.

Tufte, Edward R. *Envisioning Information*. Cheshire, CT: Graphics Press, 1990.

van Engen, John. "Authorship, Authority and Authorization: The Cases of Abbot Bernard of Clairvaux and Abbess Hildegard of Bingen." In S. Boodts et al., eds. *Shaping Authority: How Did a Person Become an Authority in Antiquity, the Middle Ages and the Renaissance?* 325–62. Turnhout: Brepols, 2016.

Vessey, Mark. "Fashions for Varro in Late Antiquity and Christian Ways with Books." In C. Humfress, C. Harrison et al., eds. *Being Christian in Late Antiquity: A Festschrift for Gillian Clark*, 253–80. Oxford: Oxford University Press, 2014.

Wagner, David. "Glimpses of Unsurveyable Maps." In R. Heinrich, E. Nemeth et al., eds., *Image and Imaging in Philosophy, Science and the Arts*, 365–76. Frankfurt: ontos Verlag, 2011.

Wallis, Faith. *Bede: The Reckoning of Time*. Liverpool: Liverpool University Press, 2004.

——. "What a Medieval Diagram Shows: A Case Study of *Computus*." *Studies in Iconography* 35 (2015): 1–40.

Walter, Katie L. *Middle English Mouths: Late Medieval Medical, Religious and Literary Traditions*. Cambridge: Cambridge University Press, 2018.

Ward, Benedicta. *Anselm of Canterbury: His Life and Legacy*. London: SPCK, 2009.

Weaver, Richard M. *The Ethics of Rhetoric*. Davis, CA: Hermagoras Press, 1985.

Wetherbee, Winthrop. *Platonism and Poetry in the Twelfth Century*. Princeton: Princeton University Press, 1972.

Whistler, Catherine et al. *Raphael: The Drawings*. Oxford: Ashmolean Museum, 2017.

White, Hugh. "Langland's Ymaginatif, Kynde and the *Benjamin Major*." *Medium Aevum* 55 (1986): 241–8.

White, Rebekah, G. Gaspar, T. McLeish et al. "Magnifying Grains of Sand, Seeds, and Blades of Grass: Optical effects in Robert Grosseteste's *De Iride* (*on the Rainbow*) (circa 1228–1230)." *Isis* 112.1 (2021): 93–107.

Williams, B. R. and R. S. "Finger Numbers in the Greco-Roman World and the Early Middle Ages." *Isis* 86.4 (1995): 587–608.

Williams, John. *The Illustrated Beatus: A Corpus of the Illustrations of the Commentary on the Apocalypse*. 5 vols. London: Harvey Miller, 1994–2003.

——. *Visions of the End in Medieval Spain*. Project Muse, 2017. https://doi.org/muse.jhu.edu/book 66670.

Williams, Shelley M. "Capturing Sound: The Art of William de Brailes in New College MS 322." *New College Notes* 17.2 (2022): 1–18.

Williams, Travis D. "Mathematical *Enargeia*: The Rhetoric of Early Modern Mathematical Notation." *Rhetorica* 34. 2 (2016): 163–211.

Wilmart, André. *Auteurs spirituels et textes dévots du moyen âge latin*. Paris: Etudes Augustiennes, 1971.

Worm, Andrea. "*Arbor autem humanum genus significat*: Trees of Genealogy and Sacred History in the Twelfth Century." In P. Salonius and A. Worm, eds. *The Tree: Symbol, Allegory, and Mnemonic Device in Medieval Art and Thought*, 35–67. Turnhout: Brepols, 2014.

Zeeman, Nicolette. *Piers Plowman and the Medieval Discourse of Desire*. Cambridge: Cambridge University Press, 2006.

Ziolkowski, Jan M. "*Amaritudo mentis*: The Archpoet's Interiorization and Exteriorization of Bitterness in Its Twelfth-Century Contexts." In C. S. Jaeger and I. Kasten, eds. *Codierungen von Emotionen im Mittelalter/ Emotions and Sensibilities in the Middle Ages*, 98–111. Berlin: De Gruyter, 2003.

Zumthor, Paul. *Essai de poétique mediévale*. Paris: Éditions du Seuil, 1972.

——. *La poesie et la voix dans la civilisation mediévale*. Paris: Presses universitaires de France, 1984.

Index

Printed and bound by CPI Group (UK) Ltd, Croydon, CR0 4YY
13/09/2026
14961006-0001